Economic Aspects of the
Energy Crisis

Economic Aspects of the Energy Crisis

Harry W. Richardson
University of Southern California

Lexington Books/Saxon House
D.C. Heath and Company
Lexington, Massachusetts
Toronto   London

Library of Congress Cataloging in Publication Data

Richardson, Harry Ward.
  Economic aspects of the energy crisis.

  Bibliography: p.
  Includes index.
  1. Power resources—United States. 2. Energy policy—Economic aspects—United States. I. Title.
  HD9502.U52R52   333.7'0973   75-8360
  ISBN 0-669-03327-8

Copyright © 1975 by D.C. Heath and Company.

All rights reserved. No part of this publication may be reproduced or transmitted in any form or by any means, electronic or mechanical, including photocopy, recording, or any information storage or retrieval system, without permission in writing from the publisher.

Published simultaneously in Canada.

Printed in the United States of America.

International Standard Book Number: 0-669-03327-8

Library of Congress Catalog Card Number: 75-8360

Contents

	List of Tables	ix
	Preface	xi
Part I	*Background and Theory*	
Chapter 1	**Trends and Projections**	3
	Energy and Economic Growth	3
	Trends	8
	Projections	15
Chapter 2	**The Resource Exhaustion Controversy**	27
	The Pessimists	27
	The Optimists	34
Chapter 3	**The Theory of Exhaustible Resources**	39
	The Profit-Growth Rule	39
	The Impact of Taxes	43
	An Environmental Damages Tax	44
	Recycling	44
	Decline in Resource Quality	45
	Technological Change	46
	Switching to a Backstop Technology	47
	The Social and Private Rates of Discount	48
Part II	*Energy Resources*	
Chapter 4	**Oil**	57
	Trends	57

	Rule of Capture	58
	Prorationing	59
	Field Unitization	59
	Taxation and Petroleum Extraction	60
	The Power of the Oil Industry	62
	The Oil Import Question	64
Chapter 5	**Coal, Gas, and Electricity**	**71**
	Coal	71
	Natural Gas	79
	Electricity	85
Chapter 6	**Nuclear Power**	**89**
	Introduction	89
	Nuclear Reactors	90
	Fusion Reaction	92
	The Backstop Technology?	93
	Nuclear Power Production	94
	Resources	96
	Health Hazards	98
	Disposal of Radioactive Wastes	100
	Nuclear Theft	101
	Conclusions	101
Chapter 7	**Other Fuels**	**103**
	Shale Oil	103
	Synthetic Fuels	104
	Solar Power	106
	Geothermal Energy	108
	Tidal Power	110
	Hydrogen	110
Part III	*Problems and Solutions*	
Chapter 8	**OPEC, the United States, and World Oil**	**115**

	Production, Consumption, and Reserves	115
	A Brief History of OPEC	117
	The Recent Course of Oil Prices	119
	The Impact of the 1973-74 Oil Embargo	121
	Understanding OPEC	123
	Strategies	129
	Emergency Stockpiles	132
Chapter 9	**Demand Restraint**	137
	Introduction	137
	Residential and Commercial	138
	Transportation	141
	Industry	142
	Instruments	143
Chapter 10	**Energy and the Environment**	151
	Pollution Impacts	151
	Environmental Implications of Power Plants	152
	Coal and the Environment	158
	Automobile Emissions	162
	Oil Spills	162
	The Trans-Alaska Pipeline Debate	163
Chapter 11	**Some Policy Considerations**	167
	Energy R and D	167
	Energy Taxes and Subsidy Policies	172
	The Federal Power Commission	174
	Policy Proposals of Three Major Studies	175
	An Energy Standard of Value?	179
Chapter 12	**Toward a National Energy Policy**	183
	Diagnosis	183
	Energy Independence	186
	Proposals	189
	Conclusion	199

Notes	201
Bibliography	213
Index	225
About the Author	235

List of Tables

1-1	Energy Consumption, GNP, and Population, 1920-70	9
1-2	Percentage Distribution of Energy Consumption by Sector and Source, 1950-70	10
1-3	US Proven Reserves	11
1-4	Foreign Trade in Fuels	14
1-5	MIT Supply and Demand Forecasts for 1980 at Different Price Levels	17
1-6	Energy Projections for Alternative World Prices and Policies	19
1-7	Growth Rates for Energy Demand by Sector, 1960-72, and Projected, 1973-85	21
1-8	Maximum 1985 Production Levels under "Business-as-Usual" and "Accelerated Supply" (at $11 Price)	22
1-9	World Nuclear Power Generating Capacity, 1973 and 1980	24
1-10	Estimates of Nuclear Power Capacity, US and Foreign (exc. Communist), 1980-2000	24
1-11	Forecasts of Gross Energy Consumption, 1980, 1985, and 2000	25
4-1	United States Oil Reserves, Production and Imports, 1950-73	58
5-1	Regional Coal Statistics	74
5-2	Natural Gas Reserves, Production and Price	81

5-3	Electric Power Industry	86
6-1	Nuclear Power Statistics	95
8-1	Crude Petroleum Production, Consumption, and Reserves by Major Producing Region, 1973	116
9-1	Household Appliances	140
9-2	Potential Energy Savings by Sector, 1985	143
10-1	Pollution and Energy Use, 1972 and 1985	152
11-1	Energy R and D Expenditures, 1963, 1973, and 1975	170

Preface

The intention of this book is to analyze the major issues arising from the so-called "energy crisis" that has dominated much of the discussion on United States economic policies since 1973. In the judgment of this observer, the severity of this "crisis" has been exaggerated by policy-makers, the media, and the general public. Although the era of cheap energy is over, and there are many unsolved energy supply and demand problems, the threat of energy resource extinction is not very serious, and there is no danger of OPEC or any other source of supply constraint bringing the American economy to its knees. The reader can make up his or her own mind on the basis of the evidence.

The book attempts to achieve three aims. First, the discussion is intended to be broad enough to provide a sufficiently deep background understanding of the major energy problems facing the United States now and in the future. Second, several recent large-scale studies of the energy situation (such as the Federal Energy Administration's *Project Independence*) that are too indigestible to be read by the nonspecialist are summarized and evaluated. Third, and perhaps most important, the book leads up to an assessment of the appropriate scope and content of national energy policy. In this highly topical field, any book will contain material that will be superseded by the pace of events. Attention is consequently focused on the underlying issues rather than on the specific aspects of short-run policy changes.

Although this book is about economics, technical as well as social and political questions cannot be ignored. These are introduced where necessary to the argument. There is no complicated economic theory, but noneconomists may prefer to skip the theoretical analysis of Chapter 3.

The book is divided into three parts. Part I deals with background and theory. Chapter 1 discusses the relationship between energy and economic growth, and presents some general background trends and forecasts on the United States energy situation. Chapter 2 poses the question of energy resource extinction by examining the arguments of pessimists and optimists on this issue, including a brief study of one of the first pessimists, the nineteenth-century economist W. Stanley Jevons. In Chapter 3 the main elements of the theory of resource exhaustion relevant to energy resources are presented. Part II looks at the recent experience, current problems, and future prospects for each major fuel—oil, gas, coal, and nuclear power—and examines the possibilities for development of other fuels at present making a negligible contribution to national energy supply. The core of the book is the analysis of problems and policy options in Part III. The problems studied include the world oil situation and relations with OPEC, demand restraint and energy conservation, the trade-off between energy supply and environmental protection, energy research and development, and the major policy issues that must be considered in evaluating the need for and scope of a national energy policy.

I am very grateful to Julie Smith, Elaine Sullivan, Esther Zavos, Sharon Hull, and Al Sciulli for typing the final draft.

Part I:
Background and Theory

1 Trends and Projections

Energy and Economic Growth

Historically, in the developed world in general and in the United States in particular, there has been a close interrelationship between economic growth, rising living standards, and increasing energy consumption. In a sense, the story of industrialization (and urbanization) is the story of how energy resources have been exploited for the use of mankind. This close association has led some observers to think of economic growth and increasing energy consumption per capita as Siamese twins. This inseparability is based upon three types of evidence: the long-run historical record and generalizations drawn from the economic history of industrialized countries; the similarity over decades of experience in the United States between annual growth rates in GNP and in energy consumption; and cross-sectional international evidence suggesting a strong relationship between level of economic development (as reflected in per capita average income) and energy consumption per capita. How far this evidence needs to be qualified is a question that can be discussed on the basis of the quality of the data and the methodology used to interpret them. What is more dangerous, however, is that many who support the hypothesis of inseparability have stepped beyond analysis of the data to derive policy conclusions.

Two opposing inferences have been made, depending upon the observer's assessment of the energy supply situation. The first is to argue that the supply of energy resources is in danger of running out, that it is a severe constraint on growth, and that in order to prevent total economic and social collapse the only way to bring energy consumption and growth into correspondence is to cut back on the pursuit of growth and hence on consumption (see pp. 31-34). The second inference is almost the opposite. It stands on the implicit assumption that energy use is one of the foundations of civilization, and that man via his control over technology will always find ways to increase energy supply. To cut back on energy consumption is not only unnecessary, it is dangerous. Measures to promote energy conservation can be implemented only at the expense of economic growth, and to risk disturbing the growth of the economy may wipe out the sources from which technological advance is possible.

Neither position is very optimistic about the prospects for breaking the nexus between energy consumption and growth. The former holds that in order to conserve energy resources, we have to abandon growth; the latter believes that if we attempt to conserve energy, growth will abandon us. Since one of the primary objectives of a national energy policy is how to maintain growth *and* conserve energy (or at least keep energy supply and demand in balance), it is important to examine the strength of the energy-growth link.

Viewed from a particular perspective, the economic history of the industrialized world might be written in terms of the increasing use of energy and the substitution of more for less productive energy-using technologies. The mechanization of industry, the succession of revolutions in transport, the improvement in material living standards and the social history of changing life styles—all these phenomena can be understood only within a context that stresses the increase in supply and transformation of energy resources. But elaboration of the historical association between economic growth and increasing energy consumption, illuminating though it may be in highlighting the reliance of modern man on fuels, does not answer the critical question of the nature of the link. Over the last century, energy consumption has increased at a rate three times faster than that of population, but GNP has increased at more than double the rate of energy use. In the twentieth century the growth rates have been much closer, but the differential has varied from decade to decade. The relationship is much more complex than appears at first sight.

The changes in energy consumption and GNP rates over time have been analyzed by Darmstadter.[1] Over the twentieth century as a whole, the parallel is close: 3.2 percent per annum in energy consumption and 3.3 percent in constant GNP. Between 1880 and 1920, however, energy consumption increased one-and-a-half times as fast; between 1920 and 1965, on the other hand, it grew more slowly, only three-quarters of the rate of GNP. The faster growth before 1920 reflected primarily the rapid growth of manufacturing. The more recent deceleration phase was due to many factors: electrification of factory operations increased fuel efficiency (despite the thermal efficiency losses in electricity generation); a doubling of thermal efficiency improvements over the 1920-65 period; and the replacement of steam by diesel engines. After 1965, the relative GNP and energy growth rates switched again. The rate of growth in energy consumption was 1.4 percent higher (5.0 as against 3.6 percent) in the second half of the 1960s compared to the first half, and was much higher than the growth rate in GNP. The three most important contributory factors were electricity conversion losses (a reversal of decades of continuous improvement), rising petroleum consumption by automobiles and planes, and increasing nonenergy uses. Outside the transportation sector, the switch to secondary energy sources in the form of electric power has been progressive, and electricity consumption has been growing much faster than energy consumption as a whole. Since the scope for major efficiency improvements in the electric power sector is probably small in the short run, dampening of demand is the main hope of reducing the rate of growth of consumption. The experience since 1973, a period of rapidly rising unit electricity prices, has been moderately encouraging.

⌈Energy is linked to economic growth in another way—the share of energy-related sectors in total output and employment. The Ford Foundation Energy Policy Project[2] estimated that the energy industry itself and energy-intensive industries (defined as having energy consumption/total output ratios more than four times the national average) consume about one-third of total energy use, and account for 45 percent of industrial production, 15 percent of GNP, but only 10 percent of employment. In the 1950s and 1960s total employment in the United States increased by 41 percent; in the energy industries the increase was only 5.5 percent, a mere 115,000 jobs. This net increase took place entirely in the utilities and the fuel marketing (especially gas service stations) subsectors, offsetting massive declines in mining, refining, and pipeline transportation.⌉

International comparisons are often used to support the view that per capita energy consumption and the level of economic development are inextricably linked. Superficially, there is a close relationship, as might be shown in a scatter diagram relating per capita consumption levels to GDP per capita. However, the generalization can be overstressed. For instance, in Sweden and Canada—countries with similar living standards and climatic conditions—energy consumption per capita in the former is only two-thirds of the level in the latter. The consumption level in the United States is more than twice that of West Germany, and three times as much as in Switzerland.

A typical procedure in international comparison is to fit a straight line of unitary slope through a log-scale scatter diagram of the type mentioned above, and use this to derive the conclusion that exponential growth in energy consumption is inevitable as other countries approach similar levels of development to those presently enjoyed by the front runners in energy consumption—the United States and Canada. There is no justification for this procedure. The logarithmic-based statistical relationship appears to narrow the distance between the energy consumption levels of North America and the rest of the world, and could result in the wrong inference that all other countries are helter-skelter on the North American development path. In fact, energy consumption levels are dependent upon the relative price of energy to other inputs, industrial structures and efficiency of energy use—variables that differ widely among countries.

Moreover, there is a degree of arbitrariness in the fitting of a scatter curve. A concave curve fits as well as a straight line, and this might be interpreted as suggesting that as countries become richer their growth rates in energy consumption will decline. Moreover, it might be more sensible to use separate curves for groups of countries at similar levels of economic development and with similar institutional frameworks. The fact is that because of differences in industrial structures and in rates of economic growth, the ratio of energy consumption to GNP growth rates has varied widely even among industrial countries. Finally, because of the worldwide shift from low to high energy prices over the last three years, it is dangerous to extrapolate from historical experience. Most projections of energy consumption rates have tended to ignore the question of price elasticities, or at best have assumed that these were very low.

These arguments suggest that it is too simplistic to assume an inevitable link between growth rates in GNP and in energy consumption.[a] The important question is whether a cutting back on the growth rate of energy consumption, due to either supply or demand constraints, will be prejudicial to the rate of economic growth. To answer this question, it is inappropriate to rely on historical experience, because no conscious attempt has been made in the past to conserve energy, at least in the context of national energy policy.

Darmstadter[3] argued, prior to recent events, that it would be difficult to break the association between GNP and energy growth, barring a major technological breakthrough or the imposition of substantial disincentives to demand. (It is unclear whether the recent rise in energy prices would qualify under this heading.) The reasons given included the close link between productivity increases and energy use; the dominance of electricity in consumption increases; the fact that the switch to services might do little to dampen energy growth because of computerization and the use in the service sector of energy-intensive technologies; and the fact that increased leisure and recreation could break either way, depending on changing tastes and life styles. However, even this study came out with a wide range of long-term energy growth projections (from 2.3 to 4.3 percent per annum), according to the choice of technological and institutional assumptions.

Both the Project Independence Blueprint (PIB)[4] and the Ford Foundation study undertook more systematic analyses. PIB used an economic-impact forecasting methodology composed of two linked models: a price-sensitive input-output table to relate energy prices to other industrial prices and to provide sectoral output predictions; and a long-run macroeconomic forecasting model to link production activity to income, the capital market, and the labor market. The Ford Foundation used an econometric forecasting model (the Hudson-Jorgenson model) that was not dissimilar: a macro-econometric model of US energy growth based on nine production sectors, and integrating both supply and demand, was linked to an input-output model in which the technical coefficients were assumed responsive to price changes. This approach avoids the implicit assumption of price inelasticity that had distorted most earlier projection models. In view of their similarities, it is not surprising that both studies came up with compatible results. Reductions in energy consumption due to energy conservation would have nonnegligible effects on economic growth, but

---

[a]For a contrary view see the comments of D.C. Burnham (in Ford Foundation Energy Policy Project, Final Report, *A Time to Choose: America's Energy Future* (Cambridge, Mass.: Ballinger, 1974), p. 367).

There is a wealth of data which substantiates the widely accepted contention that growth in energy usage and growth in the economy are inextricably linked... The historic relationship between GNP growth and energy growth has continued during the first half of 1974 despite an increase of 75 percent in energy costs and a tripling of the rate of inflation.

the costs would be small enough to warrant the conclusion that a slowing down in the growth rate of energy consumption is compatible with the maintenance of economic growth.

PIB examined the implications for growth of an accelerated supply strategy, of demand conservation, and of variations in the world price for oil. Accelerated supply would have different short- and long-run effects. In the short run, its net impact would depend on the state of the economy. In a slack economy, it would reduce unemployment and also lower inflation via relieving shortages; if the economy was overheated, on the other hand, it could raise inflation without affecting unemployment. In the long run, accelerated supply would raise the growth rate to the extent that investment in energy supply was more productive than average investment elsewhere in the economy. Conservation would, of course, reduce the demand for energy. The oil embargo showed that short-term unforeseen reductions in energy supply could adversely affect output and employment. But this offers no guidance to what might happen in the event of a planned long-term reduction in consumption. To the extent that conservation reduces waste and to the extent that substitutes are available for the conserved resources, there would be no adverse effects on real economic growth and employment. Moreover, there are clear-cut benefits in the form of less disruption to the environment and slowing down the rate of domestic resource depletion.[b]

A high world oil price would stimulate investment in energy production, but this would be more than offset by dampening effects of high fuel prices on growth elsewhere in the economy. The difference between a $7 and $11 world oil price might be in the region of 0.5 percent in the GNP growth rate. On the other hand, the unemployment rate is not very sensitive to variations in the world oil price. On the whole, therefore, PIB believes that the choice of energy policies will not have detrimental effects on the GNP. However, it admitted that there are other economic and social impacts that could be more serious: these relate to the balance of payments, inflation, regional growth, housing, and income distribution (the burden of higher energy prices falls disproportionately on the poor).

The Ford Foundation Study's basic conclusion was that its preferred strategy of energy conservation was possible without major sacrifices in real income growth and without drastic changes in the structure of the economy, even to the extent of more than halving the rate of growth of energy inputs. An energy conservation strategy of this magnitude would have the following effects: a 4 percent shortfall in real income levels by the year 2000 compared with an

---

[b]This benefit should not be exaggerated.

The only way to prevent eventual exhaustion of such a resource would be to stop totally all economic activity that used it. Limited, zero or even negative growth would only postpone the day of reckoning. Yet ending all use of the resource forever would surely be pointless, for then it might as well be gone now.

(M.J. Roberts in M. Olson and H.H. Landsberg, *The No-Growth Society* (New York: Norton, 1973), p. 120.)

extrapolated energy growth economy; more employment, but slightly lower real wages, reduced output, and lower labor productivity; slightly lower investment requirements; and a modest acceleration in the price level, raising the inflation rate by 0.3 percent.

Some of these predictions depend upon energy-saving technologies having a moderating effect. For example, impacts on producing sectors are expected to be small because of reductions in waste and because of technological improvements. Similarly, the conversion losses in electricity generation are so heavy and the share of electricity in total energy consumption so large that even modest savings in that sector due to technological improvements could have a big effect. It is debatable whether these economies will be achieved; on the other hand, given the time frame of the forecasts, they could exceed expectations.

Trends

Consumption

Key trends in energy consumption in the United States in the fifty years prior to 1970 are illustrated in Table 1-1.[c] Over this period energy consumption increased by three-and-a-half times while GNP rose by more than five times. Thus, energy consumption per GNP dollar fell between 1920 and 1965, though it ominously rose thereafter. Energy consumption per capita, on the other hand, rose from 1940 onward after two decades of stability; in the 1960s alone it increased by 30 percent. Electricity consumption was the most vigorous growth area in the energy field, increasing by almost 2900 percent over the half century and by close to 1500 percent in per capita terms. Its share of total power almost trebled over the period, reaching one-quarter of the total by 1970.

There have been significant changes in the distribution of power sources. These changes primarily occurred before 1960; since then the situation has remained relatively stable. The share of hydropower has been constant. The major shift has been a continuous decline in the share of coal—from almost 80 percent down to 20 percent of the total—offset by sharp increases in the shares of oil and gas. This massive transformation reflected shifts towards cheaper, more flexible and environmentally satisfactory fuels, technological advances

---

[c]The standard unit of energy is the British thermal unit (Btu). It is the amount of energy required to raise the temperature of one pound of water by one degree Fahrenheit. Aggregate energy demand is frequently expressed in terms of Q.s, where Q. = a quadrillion (or 10^{15}) Btus.

Conversion rates depend on the quality and type of fuel, but approximations are:

1 42-gallon barrel of oil = 5.8 million Btus
1 cubic foot of natural gas = 1031 Btus
1 kilowatt hour of electricity = 3413 Btus
1 ton of coal (high Btu) = 25 million Btus

Table 1-1
Energy Consumption, GNP, and Population, 1920-70

	1920	1930	1940	1950	1960	1965	1970
Energy cons. (tr. Btu)	19,782	22,288	23,908	34,154	44,960	53,785	68,810
Electr. cons. (b. Kwh)	57.5	116.2	182.0	390.5	848.7	1,157.4	1,648.3
GNP (b. 1958 $)	140.0	183.5	227.2	355.3	487.7	617.8	724.1
Population (m.)	106.5	123.1	132.6	152.3	180.7	194.6	205.4
Per capita							
Energy cons. (m Btu)	185.8	181.1	180.3	224.3	248.8	276.4	335.0
Electr. cons. (Kwh)	540	944	1,376	2,564	4,967	5,948	8,025
GNP ($ 1958)	1,315	1,490	1,720	2,342	2,699	3,175	3,525
Per $1 of GNP							
Energy cons. ('000 Btu)	141.3	121.5	105.2	96.1	92.2	87.1	95.0
Electr. cons (Kwh)	0.41	0.63	0.80	1.10	1.74	1.87	2.28
Energy cons. by source (%)							
Coal	78.4	61.2	52.4	37.8	23.2	23.0	20.0
Natural gas	4.2	8.8	11.4	18.0	28.2	29.9	32.8
Petroleum	13.5	26.5	32.4	39.5	44.6	43.2	43.0
Hydro and nuclear	3.9	3.5	3.8	4.7	4.0	3.9	4.1
Electricity as % of total Power	8.4	8.8	10.3	15.1	18.7	20.6	24.7

Source: U.S. Bureau of the Census, *Historical Statistics of the United States, Colonial Times to 1957* (Washington, D.C.: U.S. Government Printing Office, 1960), pp. 139, 507-8.

U.S. Bureau of the Census,*Statistical Abstract of the United States, 1974* (Washington, D.C.: U.S. Government Printing Office, 1974), pp. 5, 373, 514, 517.

Edison Electric Institute, *Statistical Yearbook of the Electricity Utility Industry for 1971* (New York, 1972).

(such as the diffusion of the automobile as the main transport mode), and changing tastes. To the extent that fuel substitutions are difficult to reverse, it also represented a shift from a highly elastic resource base to one that was much less elastic, more uncertain, and now very unstable.

As shown by the distribution data of Table 1-2, the changes in sources have been paralleled by equally significant variations in the shares of demand sectors. In the aggregate, the most striking change has been the relative decline in industrial demand and the growth of electric utilities. The more interesting developments, however, have taken place at the sectoral level: the substitution of natural gas for coal in home and office heating; the disappearance of coal in transportation with the passing of the steam train, and the even more vigorous rise in petroleum consumption for automobiles and planes; the decline in the use of coal by industry, partly offset by greater use of natural gas; and the expansion in fuel consumption by the electric power companies, particularly impressive in coal and to a lesser extent natural gas (the growth in nuclear power plants is predominantly post-1970).

Horsepower statistics (one horsepower equals 550 foot-pounds per second, or 0.7457 Kilowatt) provide another way of looking at the dramatic changes in

Table 1-2
Percentage Distribution of Energy Consumption by Sector and Source, 1950-70

		Coal	Natural Gas	Petroleum	Hydro and Nuclear	Total
Household and commercial	1950	8.5	4.8	8.9		22.2
	1960	2.3	9.5	10.9		22.7
	1970	0.6	10.7	9.2		20.5
Industrial	1950	17.4	10.9	7.7		36.1
	1960	10.9	14.0	8.2		33.1
	1970	8.1	15.3	7.4		30.7
Transportation	1950	5.0	0.4	19.9		25.2
	1960	0.2	0.8	23.1		24.1
	1970	–	1.0	22.9		23.9
Electric utilities	1950	6.5	1.9	1.9	4.7	15.1
	1960	9.5	4.0	1.3	4.0	18.7
	1970	11.4	5.8	3.3	4.1	24.7
Total	1950	37.8	18.0	39.5	4.7	100.0
	1960	23.2	28.2	44.6	4.0	100.0
	1970	20.0	32.8	43.0	4.1	100.0

Source: Federal Energy Administration, *Project Independence Blueprint* (Washington, D.C.: U.S. Government Printing Office, 1974), AI, pp. 9, 12, 15.

energy patterns in the United States in this century. In 1900, out of a total of 64.6 million horsepower, work animals accounted for more than a third (24.5 million), and two other uses—railroads (18.7 million) and on-site factory power (10.3 million)—for 45 percent of the total. The fourth most important source of horsepower was human energy itself (4.0 million). By 1970 the distribution was very different. Total horsepower in use had soared to 20,129 million, of which 96 percent (19,325 million) was due to motor vehicles alone. Central electricity generating plants accounted for 435 million and aircraft for 183 million. No other source accounted for more than 0.27 percent (railroads, on-site factory power and mines were all of that order of magnitude).

Reserves

Estimates of the United States energy reserves position are shown in Table 1-3. The fuel value of these reserves must be placed in the context of a current energy consumption of a little under 75 Q. Btu and forecasts for 1985 that converge around the 100 Q. mark. How one assesses these reserve levels depends

Table 1-3
U.S. Proven Reserves

Source	Fuel Units	Q. Btus	Years left at 1972 Consumption
Coal			
High sulfur	273 billion tons	6,908	
Low sulfur	160 billion tons	3,838	
Total	433 billion tons	10,746	823
Oil			
Lower 48 (crude)	30 billion barrels	176	
Natural gas liquids	6 billion barrels	37	
Alaska	10 billion barrels	59	
Total	46 billion barrels	272	8
Gas			
Lower 48	218 TCF	225	
Alaska	32 TCF	32	
Total	250 TCF	257	11
Shale	20-170 billion barrels	116-986	3-28
Tar sands	29 billion barrels	168	28

Source: Federal Energy Administration, *Project Independence Blueprint* (Washington, D.C.: U.S. Government Printing Office, 1974), p. 66.

on expectations of new reserve discoveries and how fast nuclear and other nonfossil fuel technologies are expected to develop. A pessimist on these two critical factors would regard only the coal reserve position as satisfactory, and might interpret the oil and gas situations as quite alarming (though offshore reserve potential is probably underestimated).

The United States and the Rest of the World

The US energy situation must be placed in its world context to be properly understood. The most striking feature is the vast differential in per capita energy consumption between the United States and the rest of the world; a ratio at present of 8:1. Although world consumption is expected to increase considerably faster than in the United States—partly because the current energy lag is so large absolutely, partly because world output is expected to grow faster than US output—and energy demand is related to output growth, the ratio of per capita consumptions is expected to narrow only slightly by the year 2000, to about 6.5:1.[5] The major reason for this is the continued vigor of world population growth.

There are some striking differences in the fuel mix. At present, both areas are equally dependent on oil. Nuclear power accounts for a trivial fraction of energy consumption in both cases. The real difference is the US reliance on natural gas, while the rest of the world is much more dependent upon coal, and to a lesser extent on hydropower. The relatively low reliance of the US on coal is perhaps surprising in view of her vast reserves, but natural gas is a much more flexible and environmentally acceptable source of power. However, present estimates of gas reserves suggest that its share must decline progressively over time.

Different parts of the world vary widely in their energy consumption levels, their production potential, and their degree of dependence on international trade. Almost 80 percent of world energy consumption is accounted for by three regions—North America, western Europe, and the USSR and the rest of eastern Europe. Asia accounts for almost 14 percent, while the shares of Latin America, Africa, and Oceania are very small at present. However, most areas consumed energy faster than the United States in the 1960s, with Japan the fastest of all. Even the low energy consumers such as Latin America and Asia consumed energy at a little faster rate than the world as a whole. In the next decade, these trends are expected to intensify—deceleration in the United States and western Europe and acceleration in Latin America, Communist Asia, and Africa. However, given the changes in world energy supplies, and *a fortiori* in prices, these growth rates in the third world may be dampened, and non-communist Asia—excluding Japan, but including the Middle East—may become the fastest-growing energy consumer.

The changes in international trade in energy products in the last two years highlight the problems of those countries which rely heavily on imports for their expanding energy consumption. Japan, western Europe, and Oceania are the most critical cases, though Latin America (excluding the Caribbean) also has problems. The difficulties of North America and eastern Europe, on the other hand, are relatively minor, though the US situation has deteriorated in recent years. The Middle East, the Caribbean, and Africa are in the strongest position, though within each of these regions the national situations vary widely. The USSR is also well placed, with a capacity for self-sufficiency despite its large size, substantial population, and rapid industrial growth.

Thus, there are wide divergences between the distribution of world production and world consumption in energy products. Given the essential nature of fuels, their relative inelasticity of demand, the heavy flows of energy products, especially petroleum, and their associated payments, attempts to balance the supply of and demand for energy at the national level could be a major contributor to international economic—and perhaps political and social—stability. On the other hand, world economic development has been promoted historically by increasing interdependence and mutual gains from trade, and the costs of a full-scale retreat into autarkic policies should not be ignored.

Foreign Trade

Table 1-4 illustrates broad trends in the United States' foreign trade in fuels. A striking characteristic is the favorable trade balance in coal (from a balance-of-payments point of view) and unfavorable balance in natural gas and petroleum products (including crude oil). Also, whereas the trade in coal has fluctuated without exhibiting an upward trend, both natural gas and oil imports have soared in recent years—particularly since 1965 in the case of gas and petroleum products and since 1970 in the case of crude oil. Of course, it is the growth of the latter that has given rise to most anxiety, particularly the growth in supplies from OPEC countries rather than from Canada, the major non-OPEC exporter to the United States. In 1970 Canada supplied over 50 percent of US imports of crude oil; by 1973 this share had slumped to 30 percent, despite the Arab embargo in the last quarter of the year. Between 1970 and 1973 Canadian crude exports to the United States increased by 50 percent, but imports from OPEC countries rose by more than 300 percent.

However, the importance of Arab oil to the United States should not be exaggerated. In 1970 imports from the four main non-Arab suppliers within OPEC (Venezuela, Nigeria, Iran, and Indonesia) were more than 50 percent greater than total Arab supplies; in 1973 they were 60 percent greater. Of the Arab countries only Saudi Arabia is a leading supplier, with 14 percent of total imports in 1973, despite the embargo. These data suggest that, despite official impressions to the contrary, it is the rising price of crude oil imports, rather than their source, that is the real problem.

Table 1-4
Foreign Trade in Fuels

Year	Natural Gas (Mcf)		Crude Oil (M. bbl.)		Petroleum Products (M. bbl.)		Coal ('000 sh. tons)	
	Imports	Exports	Imports	Exports	Imports	Exports	Imports	Exports
1950	—	25,727	178	35	133	77	365	29,360
1955	10,888	31,029	285	12	170	123	337	54,429
1960	155,646	11,332	372	3	293	71	262	37,981
1965	456,394	26,132	452	1	449	67	184	52,162
1970	820,780	69,813	483	5	765	90	36	72,411
1971	934,548	80,212	613	1	819	81	111	58,022
1972	1,019,496	78,013	811	<0.5	924	81	47	57,151
1973	1,030,000	80,000	1,178	1	1,066	85	200	53,000

Source: U.S. Bureau of Census, *Statistical Abstract of the United States, 1974*, (Washington, D.C.: U.S. Government Printing Office), p. 516.

Projections

Pitfalls

The projection of energy demand and supply is a hazardous exercise, yet one that cannot easily be ignored in view of the influence of long-run expectations on the design and formulation of national energy policy. A particular problem—highlighted by relatively recent experience—is that sudden and quite unexpected reversals in trend may occur, as in the mid-1960s when energy consumption started to grow more rapidly than GNP after more than half a century of the opposite trend. The standard approach has been to relate long-term energy growth to the growth rate in GNP, deviating around the latter according to whether growth was energy-intensive, energy-saving, or neutral. In developed countries in this century, the more usual assumption has been a slower rate of energy consumption than GNP, on the grounds that sectoral shifts in favor of service industries involve fewer energy inputs. As implied above, this assumption is of doubtful validity. For instance, in the United States in the 1960s the transportation sector was the most vigorous energy-hungry sector in terms of its rate of growth. Not only is this a service sector in itself, but its expansion was intimately associated with service sector expansion, suburbanization, and other interconnected economic and social trends.

Perhaps the most critical determinant of the rate of energy growth is what happens in the electric-power industry. Electricity generation has been increasing its share of total energy supply, due to its tendency to double every decade. Sharply rising residential and commercial demand, particularly for air-conditioning and heating, has been a major boost. More critical in the future may be the rate of productivity change in electric-generating efficiency. The heat rate,[d] which has fallen historically, began to deteriorate from the mid-1950s, worsened after 1965 (regarded as a major factor in the rising energy consumption), and is expected by most experts to decline at an even slower rate in the future. The only escape from this scenario would be a major breakthrough in electric power technology.

Even the simple hypothesis that higher incomes and sectoral shifts to services industries dampen the growth in energy demand does not stand up to scrutiny. This is because the slackening in energy consumption due to decelerating industrial growth may be offset, or more than offset, by increased automobile use, more plane trips, energy-intensive household conveniences, and other manifestations of an affluent society. Empirical studies suggest that intersectoral shifts have only minor effects on the rate of growth of energy consumption.

Projections involve other problems. For example, much depends on technological assumptions, particularly about the fuel mix. A fuel-cell economy, for

---

[d]The heat rate is the Btu requirements for generation of a kilowatt-hour of electricity.

instance, would imply much lower energy consumption than, say, an all-electric economy using power-plant technologies similar to those known today. In most scenarios the rate of substitution of electric for other forms of power and the rate of improvement in the thermal efficiency of power plants become critical variables. Economic factors must also be taken into account: the long-run price trend may have a powerful impact on the rate of growth of demand; population growth may also influence energy consumption, particularly under conditions of medium growth rates—fast enough to stimulate demand, not so fast as to divert resources to the problem of subsistence rather than development.

Moreover, the superficially neat trick of relating energy growth to GNP, even if the relationship could be specified precisely, is to escape from one projection bog merely to sink deeper in another. We have little justification for the belief that we have the power to forecast the growth rate of GNP in the long run, say up to 1990 or 2000. Indeed, it might be easier, as well as less subject to error, to adopt an "intrinsic" forecasting technique which avoids having to predict movements in "key" independent variables in favor of adjustments to past energy consumption trends.[6]

Three Projections

The MIT Energy Laboratory Policy Study Group.[7] This report prepared forecasts of energy supply and demand for 1980 on the basis of assumed prices as part of an evaluation of the costs of self-sufficiency. Although the components of the forecasts were drawn from other sources including judgmental predictions from within the domestic energy industries and the Hudson-Jorgenson econometric demand model as well as from models developed within the group, the approach is sufficiently distinctive to merit individual attention.

The key to the MIT analysis is the following argument:

When trade takes place with foreign suppliers, the price of energy can never rise above the price at which supply would equal demand if only domestic supplies were available, since importation from abroad must add some quantity to supply, and thus reduce the price. In fact, the difference between the "world trade" price and an "exclusively domestic" price is an indication of the costs—that is, the additional amounts paid per unit of consumption—that would result from invoking a policy of self-sufficiency.[8]

The study estimated supply and demand under alternative prices, expressed in terms of oil equivalent,[e] and the demand-supply gap at each of these price levels measures net imports needed to achieve equilibrium. The self-sufficiency price is the price needed to eliminate the demand-supply gap.

---
[e]A fuel is made "oil equivalent" by finding the number of barrels of oil that has the same heating value as a given quantity of that fuel.

Major uncertainties in forecasts of this kind include: the responsiveness of domestic supply and demand to price changes; the costs of developing synthetic fuels; capacity constraints in the construction industries; the world oil price; and the nature of security (US vulnerability is a complex function of the total volume of imports, the fraction of imports from any one country, and the specific sources of imports). The study used two types of forecast: judgmental—based on experienced analysis within the industry, and including qualitative as well as quantitative analysis—and econometric. The judgmental and econometric supply forecasts for 1980 were very similar, but the two demand forecasts diverged at prices above $5.50. The main reason for this is that the judgmental demand forecasts do not allow for the influence of price changes, whereas a strong demand elasticity response is built into the econometric model. Table 1-5 presents the forecast demands and supplies of the two approaches at price levels of $7, $9, and $11, and the resulting import requirements. Since low prices depress supply and (according to the econometric model) stimulate demand, low prices also imply high imports. To achieve domestic supply-demand equilibrium the price of energy has to rise to between $11 and $13 per barrel (in constant 1973 prices) under the econometric model and to over $14 according to the judgmental forecast. The MIT group argues that this implies that self-sufficiency is very costly.

Table 1-5
MIT Supply and Demand Forcasts for 1980 at Different Price Levels

		MBD equivalent at price per bl.		
		$7	$9	$11
Supply	a	36.6	38.4	38.6
	b	36.4	38.3	38.4
Demand	a	44.2	42.4	40.6
	b	45.6	45.6	45.6
Import gap	$D_a - S_a$	7.6	4.0	2.0
	$D_b - S_b$	9.2	7.3	7.2

$(D_a - S_a) = 0$ when equilibrium price (p_e) is given by $11 < p_e \leq 13$
$(D_b - S_b) = 0$ when equilibrium price is given by $p_e \geq 14$
a = econometric model
b = judgmental forecast
Source: MIT Energy Laboratory Policy Study Group, *Energy Self-Sufficiency: An Economic Evaluation* (Washington, D.C.: American Enterprise Institute, 1974), derived from tables 1 and 2, pp. 8, 9.

Much depends on what the world oil price is expected to be in 1980. If it were very high, the costs of self-sufficiency might be quite modest. However, if it settled down within the range of $7 to $11 the costs of self-sufficiency would be very high. Protection could be obtained much more cheaply via an import-storage program. For instance, the study estimates that a stockpile insuring against a one-year cut-off of 2 MBD (million barrels per day) would cost about $990 million a year, equivalent to a price increase of 25 cents per barrel or 0.67 cents per gallon of gasoline. They dispose of the argument that an artificially high price would have massive beneficial impacts on domestic supply by suggesting that current prices are high enough to stimulate oil and gas exploration, that any constraints on coal are found on the demand rather than the supply side, and that a doubling of prices might be necessary to provide a much stronger incentive for synthetic fuel development, so that this sector is better handled by specific measures, such as price guarantees for the output of synthetic fuel.

Project Independence. The FEA's forecasts refer to eight different scenarios for 1985 resulting from combinations of the following alternatives (see Table 1-6):

Business-as-usual (BAU) or accelerated supply (ACC);
With conservation or without conservation;
A world oil price of $11 or of $7.

The BAU without conservation is virtually the "no action" base case, regardless of whether the world oil price is $7 or $11. The ACC with conservation, on the other hand, implies specific actions to promote more domestic supply and to restrain demand. This strategy will be much easier to achieve at a higher world price. Accordingly, of the eight main scenarios reliance upon imports will be greatest under the BAU-without-conservation case at a world oil price of $7, and lowest under the ACC-with-conservation case at an $11 world price. The former would imply imports of 12.4 MBD in 1985, while the latter could achieve zero imports.

The implication of the PIB scenarios is that inaction will result in a serious and highly vulnerable import situation by 1985,[f] but that the introduction of policies to stimulate supply and conserve demand could, if strong enough, resolve the energy problems experienced in the United States in recent years. Since zero imports are not necessarily desirable, the most efficient socially acceptable solution probably does not call for draconian measures, though it would require some action, preferably on both the demand and supply fronts.

---

[f]PIB draws a distinction between "secure" and "insecure" imports (6.2 MBD of the latter when total imports are 12.4 MBD). However, in view of the resource conservation strategies being pursued by "friendly" suppliers such as Canada, this distinction is becoming increasingly blurred.

Table 1-6
Energy Projections for Alternative World Prices and Policies

	$7				$11			
	Base Case	Acc Supply	Cons	Combined Acc and Cons	Base Case	Acc Supply	Cons	Combined Acc and Cons
Gross consumption (Q. Btu)								
1977	82.6	n.a.	80.0	n.a.	78.9	n.a.	n.a.	n.a.
1980	91.5	91.8	86.1	86.8	86.3	87.0	82.2	82.7
1985	109.1	109.6	99.2	99.7	102.9	103.9	94.2	96.2
Domestic crude output (MBD)								
1977	7.2	n.a.	7.2	n.a.	8.0	n.a.	n.a.	n.a.
1980	8.5	9.5	8.5	9.5	10.3	11.5	10.3	11.5
1985	8.9	12.5	8.9	12.5	12.8	15.5	12.8	14.4
Oil imports (MBD)								
1977	9.2	n.a.	8.2	n.a.	6.6	n.a.	n.a.	n.a.
1980	9.7	8.4	8.1	6.8	4.6	2.9	3.3	2.0
1985	12.4	8.5	9.8	5.6	3.3	0.0	1.2	0.0
Average national Btu price ($1973/Million Btus)								
1977	1.79	n.a.	1.68	n.a.	2.24	n.a.	n.a.	n.a.
1980	1.96	1.88	1.89	1.75	2.36	2.18	2.24	2.10
1985	2.09	2.02	2.02	1.90	2.43	2.23	2.36	2.12

n.a. = not available

Source: Federal Energy Administration, *Project Independence Blueprint* (Washington, D.C.: U.S. Government Printing Office, 1974), p. 318.

The forecast imports level for 1985 with a world oil price of $11 but without strong supply stimulation or demand conservation measures is 3.3 MBD. Introducing a demand conservation strategy might save over 2 MBD of imports, while an accelerated supply strategy could eliminate imports altogether.

Some information on sectoral growth rates over the period 1960-72 and predictions for 1973-85 at PIB's two main alternative price levels (under BAU assumptions) is given in Table 1-7. Recent trends include a much faster growth in electricity consumption than in demand for primary energy resources; a stronger increase in demand for natural gas than for other fuels by all demand sectors; the declining use of coal; and a faster growth in consumption by the transportation sector than by industry.^g The forecast growth rates for 1973-85 are not particularly sensitive to assumptions about the world oil price level. Demand is expected to grow at only about one-half the rate of the recent past. The shift to electricity becomes even more intense. The household and commercial sector becomes the most vigorous demand sector, while transportation falls from top to bottom place. Coal consumption by households and commerce and by transportation continues to fall, but the industrial use of coal increases. As a result the demand for coal expands, whereas that for petroleum and natural gas slows down compared with the recent past, due to high prices and supply scarcity respectively. In fact, it is expected that household and commercial demand for petroleum and industrial use of natural gas will decline absolutely.

Maximum production potentials for 1985 as estimated by PIB are shown in Table 1-8. These estimates suggest that only the major energy sources—oil, gas, coal, and nuclear power—will make a major contribution to supply by that date. However, for these major fuels a strategy of stimulating domestic supplies could make a substantial difference, expanding oil production by a third, gas by a quarter, and nuclear power by a fifth, over the BAU output predictions. The greatest scope is in coal production, where an accelerated supply strategy could almost double expected 1985 output. However, such a strategy would involve hard choices unless desulfurization was completely and cheaply solved: either the environmental implications of a rapid development of western coal, or relaxation of air-pollution standards. Although an accelerated supply strategy could make a substantial impact on the rate of production of other fuels, such as synthetics, oil shale, and geothermal and solar power, none of these would be on a scale to be a major influence in the energy supply position by 1985. Nevertheless, the expansion in the major energy resources could be sufficient to

---

^gIn the 1960s the fastest-growing types of activities from the point of view of energy use were air-conditioning, lighting and miscellaneous uses, and household refrigeration—all relatively minor uses in terms of demand. The four main activities (transportation, process steam, direct heat in industry, and residential space heating), accounting for 64 percent of demand, grew more slowly than the average. (Office of Science and Technology, *Patterns of Energy Consumption in the United States* (Washington, D.C.: U.S. Government Printing Office, 1972), pp. 6-7.)

Table 1-7
Growth Rates for Energy Demand by Sector, 1960-72 and Projected, 1973-85 (Percent per annum)

1960-72	Coal	Petroleum	Natural Gas	Total Fossil Fuel	Electricity Distributed	Net Demand
Household and commercial	−7.5	2.6	5.0	3.1	8.8	3.9
Industry	−0.8	3.7	4.4	2.8	5.5	3.1
Transport	−21.7	4.0	6.8	4.1	−0.5	4.1
1973-85 ($7)						
Household and commercial	−8.8	−0.04	2.7	1.4	7.3	2.8
Industry	3.7	3.7	−0.9	1.6	4.7	2.0
Transport	−14.7	1.7	3.6	1.8	6.3	1.8
1973-85 ($11)						
Household and commercial	−8.2	−1.6	2.7	0.8	7.1	2.4
Industry	3.4	1.7	−0.2	1.2	3.6	1.5
Transport	−14.7	0.7	3.6	0.9	6.3	0.9

Source: Federal Energy Administration, *Project Independence Blueprint* (Washington, D.C.: U.S. Government Printing Office, 1974), AII, pp. 71, 78.

Table 1-8
Maximum 1985 Production Levels Under "Business as Usual" and "Accelerated Supply" (at $11 Price)

Source	Business as Usual	Accelerated Supply
Oil	15.0 MBD	20.0 MBD
Natural gas	23.4 TCF/year	29.3 TCF/year
Coal	1.1 billion tons/year	2.1 billion tons/year
Nuclear	234 million Kw	275 million Kw
Coal gasification	0.5 TCF/year	1.0 TCF/year
Coal liquefaction	0	0.5 MBD
Shale oil	0.25 MBD	1.0 MBD
Geothermal	6000 megawatts	15,000 megawatts
Solar heating and cooling	0.3 Q. Btus	0.6 Q. Btus
Solar electricity	41 million Mwh/year	151 million Mwh/year

Source: Federal Energy Administration, *Project Independence Blueprint* (Washington, D.C.: U.S. Government Printing Office, 1974), p. 67.

eliminate most, or even all, imports, if that were considered a sensible policy goal.

Apart from doubts about the desirability of such an objective, increasing domestic energy supplies would require overcoming other constraints, in addition to tough and costly promotion policies. For instance, decisions would have to be taken well in advance because of the substantial lead-times: twenty years for a hydroelectric plant, eight or nine years for a nuclear power plant, six years for shale oil plants and underground coal mines on federal lands, and five years for other mines, and synthetic and geothermal plants, with only oil fields having a relatively short lead-time (one to three years onshore, and two to four years offshore). Also, the capital requirements of expansion in energy supply would be very heavy. Estimates of these requirements for 1975-85 vary between $380 billion and $476 billion (at 1973 prices). The latter is the Federal Energy Administration's estimate, and is larger than comparable estimates by A.D. Little, the National Academy of Sciences, and the National Petroleum Council. Individual items vary more widely than the overall cost estimates, but the studies are in broad agreement that oil and gas, electricity generation and transmission, and nuclear power will absorb most of the capital ($188, $143, and $105 billion respectively, according to the FEA), with negligible—by comparison—amounts for coal, synthetics, and solar and geothermal projects. Transportation would require substantial capital, however, particularly to provide the means for delivering western coal to eastern markets.

The Ford Foundation Study. This explores three different scenarios, labeled "historical growth" (HG), "technical fix" (TF), and "zero energy growth"

(ZEG). The difference between them lies primarily in the projected rates of energy consumption. The HG scenario assumed a continuation of the 1950-70 growth rate in energy use, 3.4 percent per annum, to the year 2000. The TF scenario implies a slightly lower rate of economic growth, and reflects conscious attempts to conserve energy via the use of technology; the result is an energy growth rate of 1.9 percent per annum. The ZEG scenario permits the same GNP as the TF case for the year 2000, and higher employment. However, it implies additional emphasis on energy conservation, perhaps via an energy excise tax, on top of the use of energy-saving devices. It assumes a modest but declining rate of growth in energy use, reaching zero before 1990.

The three scenarios have much in common. They are not predictions per se, but illustrative possibilities to aid comparison of the effects of alternative policy choices. Each allows for enough energy to cater to basic needs such as protection against climate and the ownership of cars and other domestic appliances. Each is compatible with full employment and steady growth in GNP and in personal incomes; real GNP, for instance, is approximately doubled by the year 2000 in all three cases.

The Ford Foundation study rejects the HG scenario as consuming too many energy resources and running high import dependence risks, would favor ZEG but believes the adjustments needed would be too severe at present (though it might become a practicable goal by the late 1980s), and hence comes out in support of the TF scenario as an economically, politically, and socially acceptable and feasible strategy. However, attainment of the TF projections will not be easy since, as implied above, this requires cutting the future rate of growth in energy consumption to almost one-half of the rate experienced in the recent past.

US and World Nuclear Power Projections

Some forecasts of nuclear-power generating capacity in the United States and elsewhere are summarized in Tables 1-9 and 1-10. The predictions up to 1980 are subject to less error than the longer-run forecasts up to 2000, because plants coming into production in 1980 are already in the planning or construction phase because of the long lead-times. At present the United States accounts for three-fifths of nuclear capacity, the UK for 10 percent, and the USSR and France for 5 percent each. By 1980 the situation will have changed quite dramatically. The US dominance will remain, with about 45 percent of world capacity, but Japan will have become the second biggest nuclear-power producer, with more than 10 percent of capacity. West Germany, Sweden, and South Africa will emerge as major producers, while the UK (mainly because of the small size of her plants) and France will have slipped. The predictions up to the year 2000 exclude the Communist world. They are, of course, highly speculative, but it is possible that capacity could increase tenfold between 1980

Table 1-9
World Nuclear Power Generating Capacity, 1973 and 1980

	1973		1980	
Country	No. of Plants	Capacity Mw	No. of Plants	Capacity Mw
United States	50	31,100	149	131,200
Japan	5	1,756	41	31,636
W. Germany	7	2,082	22	14,995
United Kingdom	28	5,335	43	14,479
USSR	10	2,457	23	11,997
Sweden	1	440	13	10,060
France	6	2,481	12	7,281
S. Africa	–	–	8	6,898
Canada	5	1,974	9	5,482
Others	15	3,553	63	58,172
Total	117	51,178	380	292,200

Source: M. Willrich and T.B. Taylor, *Nuclear Theft: Risks and Safeguards* (Cambridge, Mass.: Ballinger, 1974), Appendix B, Tables B2 and B3, pp. 196-7.

Table 1-10
Estimates of Nuclear Power Capacity, US and Foreign (exc. Communist) 1980-2000

	United States Capacity (Gw)				Foreign Capacity (Gw)	
Year	Total	LMFBR			Total	LMFBR
		H	M	L		
1980	131	0.5	0.5	0.5	140	0
1985	280	9	0.5	0.5	304	1
1990	510	101	7.9	0.5	578	21
1995	810	364	99	0.5	970	106
2000	1,200	750	412	0.5	1,457	337

Source: M. Willrich and T.B. Taylor, *Nuclear Theft: Risks and Safeguards* (Cambridge, Mass.: Ballinger, 1974), Table 4.2, p. 63, and Appendix B, Table B4, p. 198.

and 2000, both in the United States and elsewhere. The prospects for nuclear power will be conditioned by the date of entry, if at all, of the breeder reactor. Assuming that the liquid-metal fast breeder reactor (LMFBR) is the most probable technological choice, its share of total nuclear capacity could vary anywhere between negligible and more than 60 percent, depending upon how quickly its practical problems are solved.

Conclusions

The dangers of energy projections are considerable, and increase with the projection period. Also, forecasters tend to be influenced by recent experience. (For instance, as the possibility of deceleration in energy consumption growth has become apparent in the last two years or so, more recent projections for 1980 and 1985 tend to be lower than those prepared in 1972.) There are so many uncertainties when the analyst looks beyond a decade that the longer-run projections can be no more than modified extrapolations. These uncertainties include doubts about the availability of resources (especially oil and gas), the difficulties of predicting new technologies and their adoption rates in energy supply, transportation and consumption, changing life styles and their implications for energy use, and the effects of energy policies still to be developed. (PIB predicts the following take-off dates: in supply technology, 1985 for *in situ* shale processes, 1990 for coal synthetics advanced systems, solar conversion, and geothermal sources, 1990-2000 for breeder reactors and 2000 for magnetic and laser fusion; in consumption, 1990 for electric or hydrogen cars and solar total energy systems; and in generation and transmission, 1990 for high efficiency closed and combined cycles for electricity generation and for cryogenic or superconducting transmission systems and 2000 for large-scale electrolytic or thermal production of hydrogen.) However, even the shorter-term projections vary considerably. (Some representative examples including those

Table 1-11
Forecasts of Gross Energy Consumption, 1980, 1985, and 2000 (Quadrillion BTU)

	1980	1985	2000
Project Independence: base case	86.3	102.9	147
conservation—major shift	82.2	94.2	120
Ford Foundation: historical growth	–	116.1	186.7
technical fix	–	91.3	124.0
National Petroleum Council: medium case	92.8	108.8	–
MIT: econometric model	97.4	–	–
judgmental model	83.4	–	–
United States Department of Interior (Dupree-West)	96.0	116.6	191.9
National Academy of Engineering	–	104.2	–

Sources: Federal Energy Administration, *Project Independence Blueprint* (Washington, D.C.: U.S. Government Printing Office, 1974), p. 431, AI, p. 17.
Ford Foundation Energy Policy Project, *A Time to Choose: America's Energy Future* (Cambridge, Mass.: Ballinger, 1974), pp. 21, 46.
W.A. Dupree and J.A. West, *United States Energy Through the Year 2000* (U.S. Department of the Interior, December 1972).

discussed above are given in Table 1-11.) It is not easy to select the most likely consumption levels from among these forecasts, and each study usually prepared a range of forecasts depending upon assumptions about supply, energy policy, prices and other variables. The dimensions to these variables suggested elsewhere in this book probably imply a consumption of under 85 Q. in 1980, 100-105 Q. in 1985, and—entirely speculative—140-50 Q. by the year 2000. These are informed guesses rather than hard projections or analytic predictions.

2 The Resource Exhaustion Controversy

The Pessimists

Jevons on Coal

W. Stanley Jevons was the first major economist to cry wolf on energy. His views, expressed in his book *The Coal Question*,[1] are worth a reexamination today, for several reasons. His arguments have many similarities with such current analyses as *The Limits to Growth*[2]; his thesis that energy is the ultimate constraint on growth is now widely accepted; if his assertions (e.g. on the commercial impossibility of substitutes) are interpreted as assumptions, his model of the contradictions between an exhaustible resource and a continuous growth in its demand is unassailable; and the weaknesses in his analysis (reliance upon extrapolation as forecasting methodology, the impossibility of foreseeing the future, overstatement of his case, the lack of or infeasible policy prescriptions) are the same found in the current literature. Moreover, *The Coal Question* is a masterpiece of how to blend popular writing with scientific analysis: well researched, well written, interdisciplinary, and to his Victorian contemporaries startling and counterintuitive.

The focus of Jevons's analysis is not world energy supplies, but the competitiveness and industrial supremacy of the British economy. He argues that this supremacy was based, both directly and indirectly but solely, on cheap and plentiful supplies of coal. Coal

> is the material source of the energy of the country—the universal aid—the factor in everything we do. With coal almost any feat is possible or easy; without it we are thrown back into the laborious poverty of early times.[3]

Despite the exaggeration, Jevons's argument here antedates input-output analysis by stressing the importance of coal in the production and distribution of Britain's export commodities.

The problem with this dependence is that coal is a nonreproducible resource:

> While other countries mostly subsist upon the annual and ceaseless income of the harvest, we are drawing more and more upon a capital which yields no annual interest, but once turned to light and heat and motive power, is gone for ever into space.[4]

The danger is more imminent than appears at first sight, because "we must discriminate the physical and commercial possibilities. The second presupposes the first, but does not follow from it."[5] The source of trouble is not the fear of physical exhaustion of resources but rather that the higher costs associated with resort to deeper, less accessible and thinner coal seams will inevitably slow down growth. "It is by this rise of price that gradual exhaustion will be manifested."[6]

However, the finite resource extractable only at increasing costs has a major accomplice—the rate of growth of demand for coal: "The exact quantity of coal existing is a less important point than the rate at which our consumption increases."[7] This is determined by the rate of growth of population and the rate of growth in coal consumption per capita (in turn, largely dependent upon the rate of growth in output). Drawing upon the Malthusian population principle, "Living beings of the same nature and in the same circumstances multiply in the same geometrical ratio."[8] Also, "Some of the main branches of industry depending upon the use of coal have obeyed the law of uniform geometrical increase for long terms of years,"[9] another illustration of "the natural law of growth, or multiplication of social affairs."[10] Thus, "The consumption of coal will increase at a nearly constant rate until some check, some natural but perhaps elastic boundary of our efforts, is encountered."[11]

As a quantitative illustration, Jevons assumes that demand for coal in the United Kingdom would increase at a rate of 3.5 percent per annum, an extrapolation of performance prior to 1861, beyond its 1861 level of 83.6 million tons. This generates a 1961 consumption level of 2607.5 million tons. The actual level was 190.5 million tons, but in addition net imports of petroleum products amounted to about 53 million tons in 1961, while gas consumption and electricity generation, both using coal—but more efficiently— amounted to 2683 million therms and 128 billion kilowatt-hours respectively. On his assumptions, aggregate coal consumption over the period 1861-1970 would have amounted to 102.7 billion tons, compared to the best available contemporary estimate of reserves of 83 billion tons down to a depth of 4000 feet, more than 1500 feet deeper than the deepest mines of the 1860s.[12] Jevons argues that if all coal had to be brought up from an average depth of 2000 feet the price of coal would have to double; the cost of coal mining at 4000 feet would be much higher, e.g. shaft expenditures increase more than proportionately with depth. The consequences are clear:

We cannot long maintain our present rate of increase of consumption.... We can never advance to the higher amounts of consumption supposed.... The check to our progress must become perceptible within a century from the present time.... The cost of fuel must rise, perhaps within a lifetime, to a rate injurious to our commercial and manufacturing supremacy; and the conclusion is inevitable, that our present happy progressive condition is a thing of limited duration.[13]

It is important to stress that the analysis does not depend substantively on the estimate of coal reserves or on the assumed rate of growth of demand. The editor of the third edition, A.W. Flux, writing in 1906, uses new estimates of reserves of 140 billion tons and a much lower rate of growth of demand, only 2 percent per annum. On these assumptions, one half of the reserves would have been exhausted by 2004. The quantity of *economically* recoverable reserves sets a ceiling on living standards while the rate of growth of demand determines the terminal date, but the argument and its inevitable conclusion remain unaffected.

Are there any exit routes? According to Jevons, the answer is No. Improvements in efficiency in the use of coal may make matters worse, because "economy of fuel leads to a great increase of consumption."[14] In other words, the price elasticity of demand for coal is high. Attempts to place a ceiling on production are either ineffective or dangerous, for

A sudden check to the expansion of our supply would be the very manifestation of exhaustion we dread. It would at once bring on us the rising price, the transference of industry, and the general reverse of prosperity, which we may hope not to witness in our days.[15]

An export duty on coal would only shift the burden to the British shipping industry. Reliance on imports is not a satisfactory solution because of high freight costs, and the resultant effect of higher coal costs feeding into export prices.

In hindsight, the introduction of substitute fuels and more efficient secondary energy sources from coal has provided at least a temporary solution. Jevons underestimated the possibilities, though his views in different parts of the book are somewhat contradictory. "Of course, it is useless to think of substituting any other kind of fuel for coal,"[16] and "The sudden demand for the manufacture of petroleum, added to the steady and rising demand of the gas works, will use up the peculiar and finest beds of oil and gas-making coals in a very brief period."[17] Later, he is a little more optimistic: "I do not deny that if our coal were gone, or nearly so, and of high price, we might find wind, water, or tidal mills, a profitable substitute for coal,"[18] and "It is just possible that some day the sunbeams may be collected, or that some source of energy now unknown may be detected."[19] However, these scientific possibilities are of no interest to Jevons, because of his concern with the "progress of the nation"; Britain would have no competitive advantage in alternative sources of energy.

Jevons offers no solution. Paying off the national debt is suggested as a palliative to compensate posterity, "who will undoubtedly suffer from an increased price of coal, the worst of taxes,"[20] for the present generation's lavish use of cheap coal. (This remedy is inappropriate on two counts. First, the debt cannot be an intergenerational burden except under very restrictive assumptions. Second, the transfer from taxpayers to bondholders is more likely to accelerate growth—and coal consumption—via a higher rate of savings.) The only choice is "between brief but true greatness and longer continued mediocrity,"[21] essentially the same options, though with a different slant, offered by the zero-growth school today. Jevons's preferred course is not to cut back on growth, but rather to use the fruits of growth wisely—so as to be able to enjoy the social and moral benefits of J.S. Mill's stationary state. Jevons explores the counterfactual world of whether Britain should have chosen not to grow in the past. His answer is clear:

in fearlessly following our instincts of rapid growth we may rear a fabric of varied civilization, we may develop talents and virtues, and propagate influences which could never have resulted from slow restrained growth however prolonged.[22]

Incidentally, Jevons provides a strong justification for futurology.

The thoughtless and selfish, indeed, who fear any interference with the enjoyment of the present, will be apt to stigmatize all reasoning about the future as absurd and chimerical. But the opinions of such are closely guided by their wishes. It is true that at the best we see dimly into the future, but those who acknowledge their duty to posterity will feel impelled to use their foresight upon what facts and guiding principles we do possess.[23]

Could the Club of Rome or the Hudson Institute have said it better?

To summarize Jevons's arguments largely in his own words has the considerable advantage that despite the archaic language, it shows how modern his analysis is. Readers of today's apocalyptic literature in the energy and general environmental fields will recognize many familiar chords: the use of research results in the natural sciences; the reliance on extrapolation using exponential trends; the significance of the absolute constraint on growth; the overstatement of the case; the impossibility of allowing for unforeseen future events that abort or postpone the predictions; the emphasis on so-called "natural laws"; the pessimism; and the absence of practical policy proposals, as opposed to demands for dramatic changes in social attitudes. In fact, the Jevons model is the kernel of most analyses on this theme published today. Moreover, judged as a model rather than as a forecast, it is correct. The role of models in this area is to tell us what *could* happen, not what *will* happen. Furthermore, though it is not possible here to evaluate the contribution of dwindling or more costly energy resources, Jevons was quite right in his prediction of the end of Britain's industrial supremacy.

For the economist, the most interesting finding of Jevons's analysis is that the real problem is one of rising costs of energy rather than of physical exhaustion of supplies. This important argument is continually being rediscovered. As a recent example, consider:

The clear evidence is that the future will not be limited by sheer availability of important materials; rather, any drag on economic growth will arise from increases in costs.[24]

Energy in the Limits to Growth *Model*

Although the MIT Project on the Predicament of Mankind, sponsored by the Club of Rome, the results of the first phase of which were published as *The Limits to Growth*,[25] was not begun until more than one hundred years after Jevons's *The Coal Question*, it uses a remarkably similar analytical approach. In both cases, the key feature is that growth collapses because an exponentially increasing population and industrial output per capita hit against a finite stock of nonrenewable resources. The differences in the *Limits* model are merely the reliance on a system dynamics model incorporating many positive feedback loops; a focus on the world rather than on one country; the treatment of energy resources as a subset of a broader category of nonrenewable resources; allowance for additional constraints on growth such as food supply, nonenergy depletable resources, and pollution; and the fact that, in the main model, the resources constraint operates via physical exhaustion rather than by the effects of rising prices. These refinements do not alter the basic feature of the incompatibility between exponential growth and a finite world (whether of resources, food, or the carrying capacity of environmental media).

Data examined by the *Limits* group for the three major fossil fuels (coal, natural gas, and petroleum) and for most metals suggested that in each case growth of consumption is exponential; the exhaustion date is not far off unless consumption is drastically cut, even if reserves are assumed to be *five* times known levels; the distribution of reserves, production, and consumption over the world is very unequal with the danger "that the political question may arise long before the ultimate economic one."[26] (The study makes use of different ratios of reserves to consumption. The static index is the reserves/current consumption ratio. The exponential index is the number of years before resource exhaustion assuming that consumption continues to grow at an intermediate growth rate (the middle of three), while a third index measures the years of consumption if reserves were five times greater than known levels. Because of exponential growth, increases in the resource stock have a proportionately much smaller impact on the exhaustion date. This supports the point made by Jevons that the rate of growth of consumption is more critical than the level of resources.) These findings apply to all nonrenewable natural resources, not only to energy resources. A more complex resource model taking into account varying quality, production costs, new technology, price elasticity of demand, and substitution does not alter the results in substance. The typical time path is: initially rapid consumption coupled with a high depletion rate and low prices (due to new technology); a second phase of rising prices, becoming cumulative as technological advances fail to keep pace with the rising costs of discovery, extraction, processing, and distribution; higher prices, inducing more efficient use of the resource and substitution of other resources wherever possible. Finally, prices become prohibitively high and production stops, frequently before the stock is completely depleted.

Drawing upon empirical evidence, the *Limits* model represents resource usage as a S-shaped curve relating resource demand (per capita resource usage rate) to wealth (industrial output per capita). In the standard model run, i.e., assuming no major change in the current dynamics of the system, it is nonrenewable resource depletion that brings about collapse.

The standard model run interpreted resource availability "optimistically" as a 250-year supply of all resources at 1970 usage rates. If the resource constraint is relaxed even further by assuming a static index of 500, the result is still an end to growth, except in this case the proximate cause of collapse is a massive increase in the level of pollution, due to an overloading of the natural absorptive capacity of the environment (though even in this case resources are severely depleted). Among the major factors in pollution are those due to energy consumption, and the *Limits* study explicitly discusses three: release of carbon dioxide into the atmosphere, the climatic effects of thermal pollution, and the increase in the level of radioactive wastes.[27] In all three cases, the evidence supports the hypothesis of exponential growth.

Other runs of the model assume that resources are "unlimited," due to the advent of fast breeder reactors and perhaps even fusion nuclear reactors, permitting better recovery of inaccessible resources and making recycling substitution possible. Once again, the system ends in collapse due to industrial pollution. However, on the assumption that nuclear energy would enable industrial pollution to be controlled technologically, a further run allows for a reduction in pollution generation from all sources by a factor of four starting in 1975. On this occasion, neither resource depletion nor pollution becomes a serious problem, but growth stops because of a food shortage.[a] To prevent growth and collapse, the only solutions in the *Limits* model are to stop exponential growth in population and output by setting births equal to deaths and industrial capital investment equal to capital depreciation; to introduce these changes sooner rather than later (the year 2000 is too late); and to combine them with other growth-control policies (reduction in resource consumption, shifts from manufacturing to services, pollution controls, more investment in food production, improvement in agricultural fertility, and technological advances to raise the average life of industrial capital). This is defined as a "state of global equilibrium," in which "population and capital are essentially stable."[28] However, "technological advances should be both necessary and welcome in the equilibrium state,"[29] including the long-term remedy of harnessing solar energy, the most pollution-free power source.

The general criticisms of the *Limits* model are too well known to need much

---

[a]Allowing for increased agricultural productivity and "perfect" birth control plus relaxing the resource and pollution constraints also fail to prevent the end of growth. "The basic behavior mode of the world system is exponential growth of population and capital, followed by collapse" (D.H. Meadows et al., *The Limits to Growth* [New York: Universe Books, 1972], p. 42). The major problem is that there are time delays in "the web of interlocking feedback loops" that hold back readjustment in a system that is itself changing rapidly.

emphasis: doubts about whether the relationships of the model reasonably represent real world phenomena, because of inadequate data and assumptions about the relationships that neglect economics and sociology; its strong neo-Malthusian content; possible underestimation of the scope for technological progress; simplistic use of approximate and very specific types of data to derive general relationships; too much emphasis on physical rather than on political and social limits; underestimation of the adaptive response of human beings; and arrogation of the results of mathematical models and system dynamics to the status of prediction, however conditional, whereas the "validity of any computer calculations depends entirely on the quality of the data and the assumptions (mental models) which are fed into it."[30]

From the point of view of depletable natural resources, and energy resources in particular, there are other problems with the *Limits* model. Its assumptions depend far too much on assertion rather than on scientific prediction. The group's response that the assumptions do not make much difference, and testing this by substituting a more optimistic set, is unsatisfactory; the kind of modification required is in the structure of the model, not in the assumptions within a given structure. For example, the analysis of the nonrenewable resources subsector depends on the assumptions of a finite and known stock of resources and a continuation in the current *rate* of growth of resource use; if these are relaxed, the pollution impacts come into play. But a reexamination of the earlier MIT model (World 2;[31] the *Limits* model is World 3) showed that overshoot and collapse did not occur (1) if the assumption that the cost of resource extraction depends on the amount already extracted is substituted for the finite stock assumption, and if capital is allocated to resource extraction as costs rise; and (2) if capital is allocated to pollution control whenever pollution levels rise.[32] These suggestions, regardless of their precise merits, illustrate the potential significance of adaptive processes, undoubtedly underrated in the *Limits* models.

In fact, the geological possibilities are not a constraint on resource availability.[b] What counts is technological and economic feasibility. The impact of market conditions on the rate and direction of resources technology, consumption, and availability is severely underestimated in the *Limits* model. For instance, the assumption of exponential energy growth is based on a statistically ambiguous interpretation of past consumption trends, particularly in a period when it is well known that consumption has been stimulated by very low prices. It is at least possible that the consumption of existing fossil fuels may be moderated by rising prices, substitution of new energy sources, and the introduction of stronger energy-conservation policies. True, rapidly rising prices

---

[b]"The model's assumptions put us much nearer the point of resource exhaustion ... than is geologically the case, and this is the main reason for the stubborn tendency of the World 3 model to collapse" (H.S.D. Cole et al., *Models of Doom* [New York: Universe Books, 1973], p. 37).

would create enormous problems of economic and social readjustment for poor, particularly resource-starved, countries by holding back the development of other sectors; but that is a different question. Similarly, the scope for technological improvements in exploration, extraction, and processing could be much greater than implied in the *Limits* model. Its assumption that it is limited rests on assertion rather than examination of either the historical evidence or a systematic assessment of current scientific opinion. If it could be assumed, for instance, that the *sum* of the annual rates of increase in resource discovery, recycling, and economy of use added up to more than 2 percent, resource collapse in World 3 is avoided.

The analysis of the pollution subsystem is weakened considerably by aggregating all types of pollution into one category, and by ignoring the fact that much pollution is local rather than global. The links between energy consumption and pollution are not spelt out.

Apart from the fact that energy is said to be related to thermal release and particle production from fossil fuels, nothing is said about how any of these might relate to pollution.... This coupling of the two "pollution" sources from energy in the argument appears to have been done so that no escape is allowed—if nuclear power is substituted for fossil fuel to escape particle production then the lower efficiency of the former can be said to result in more "thermal" pollution.[33]

These qualifications to the analysis of energy resources in the *Limits* model are not intended to show that its predictions are wrong. The truth is that no one knows. However, the use of a computer model and the repetition of countless runs illustrating the "overshoot and collapse" behavior mode may dull the critical sensitivity of the reader to the possibility that its results may be spurious. Also, the clarion call for abstinence, strong action, and changing social values *now* tends to make the skeptic feel like a heel or, worse, a traitor to the rest of humanity. The inputs into the model are certainly not beyond challenge. The *Limits* exercise illustrates very well the speculative, and in many cases value-loaded, hypotheses that are fed into apocalyptic models, whether "computer" or "mental" models. Leaving aside the jargon and the computer tricks, the analysis of the *Limits* group goes little further than—and in the discussion of economic relationships such as the influence of price on demand or the possibilities of substitution not nearly as far as—the study by Jevons, one man researching in the British Museum, more than a century before.

The Optimists

Barnett and Morse

Although the pessimists have dominated the resource-scarcity controversy, no doubt reflecting the dominating influence of the classical economists on the evolution of economic thinking, there is a contrary view—usually held by economists with a strong belief in the potential for technical progress. No one has expressed this argument more forcibly than Barnett and Morse.[34] Although their analysis refers to extractive resources in general, it can be applied to the particular case of energy resources.

They start by drawing a distinction between Malthusian scarcity (the concept of a fixed, unalterable resource supply) and Ricardian scarcity (resource limitations are economic not physical; i.e., the stock of resources is expandable, but only under conditions of declining economic quality). In the latter case it may be possible to avoid resource constraints if the resort to poorer quality resource costs can be offset by improved extraction methods and, much more important since it can lead to the substitution of entirely new resources, by technological progress. Thus, natural-resource scarcity need not imply diminishing returns to the production process as a whole.

The argument is that the Malthusian scarcity concept is irrelevant in a dynamic world, and could only apply, if at all, in primitive and isolated societies.

Recognition of the possibility of technological progress clearly cuts the ground from under the concept of Malthusian scarcity. Resources can only be defined in terms of known technology.... The notion of an absolute limit to natural resource availability is untenable when the definition of resources changes drastically and unpredictably over time.[35]

Barnett and Morse present two indirect tests of the Ricardian scarcity hypothesis in a world of technical progress. The strong hypothesis is that the economic quality of resources declines despite technical progress. Since changes in the economic quality of resources cannot be measured directly, changes in the cost of extractive output are used as a proxy. With the exception of forestry, labor-capital input per unit of extractive output was halved over the period 1870-1957. In the case of minerals, unit costs fell to a quarter of their nineteenth-century levels. Thus, the evidence suggests increasing rather than diminishing returns. The weaker hypothesis is that declining resource quality partly offsets technical advances. In this case, unit costs of extractive goods would have risen relatively to those of nonextractive goods. Again, forestry provides the only support for the hypothesis. The relative costs of total extractive output and of agricultural goods remained constant, while those of minerals were halved. Although costs for individual extractive products may rise, cost-reducing innovations and/or the development of substitutes in response to increasing costs ensure that rising costs for extractive output as a whole are avoided, or at least this is the record up to now.

Moreover, salvation through technical progress is not the result of chance or whim. Technical progress is an integral part of the development process. It takes place in response to signals, sometimes political and social, more frequently the signals of market forces. It is improper to ask what would have happened to the economic quality of natural resources in the absence of technological progress, because "the technological progress that has occurred was a necessary condition for the growth that has occurred, and if the former is ruled out the latter cannot appropriately be taken as a given fact."[36]

There are two major inferences from this analysis. First, resource scarcity can be avoided by technological change, especially by resource substitution made feasible by new technology. "Nature imposes particular scarcities, not an inescapable general scarcity."[37] Second, a consequential implication is that conserving resources for future generations may be a disservice to our descendants. Higher production today, if it means more research and investment, is more valuable to future generations than resource conservation and lower current output. The inheritance passed on is richer in terms of knowledge, technology, and institutions than if these had been sacrificed to preserve a stock of a possibly obsolescent resource.

> By devoting itself to improving the lot of the living, therefore, each generation, whether recognizing a future-oriented obligation to do so or not, transmits a more productive world to those who follow.[38]

These last arguments rely heavily on either extrapolation or faith. They are much more contentious than the distinction between Malthusian and Ricardian scarcity or the empirical analysis of resource prices over time. They provide a useful corrective to some of the wilder statements of the pessimist school. However, some analysts have argued that resource prices have moved sharply upward relative to other goods and services since the end of the Barnett and Morse study period. Minerals data on price/labor ratios show that only copper fits this argument; for ten other minerals, including coal and petroleum, these ratios fell substantially in the 1960s.[39] Moreover, the determinants of the recent rise in crude petroleum prices have nothing to do with resource scarcity or rising extraction costs. At the same time, Barnett and Morse's findings applied to highly aggregated commodity groups. In the energy-resources field, there are few substitution possibilities in the very near future. But the price signals have not had long in which to exert their effect, and it is not yet clear how persistent they will be.

Nordhaus

A more recent expression of the optimistic case is that of Nordhaus.[40] He accepts the need for more care in the use of energy resources than in the past and for more consideration to the environmental impacts of rising energy consumption, summarized in the plea for a transition from a "cowboy" to a "spaceship" economy. Nevertheless, his views on energy-resource scarcities are decidedly optimistic. He dismisses the global models as being "theoretical" and capable of generating opposite results with a different theoretical structure.

Second, he points out the pessimistic bias in evaluations of resource availability in terms of R/C ratios (i.e. the ratio of proved reserves to current consumption). The more important concept is that of URR (ultimate recoverable resources), and this is a function of technology and price. URR/C ratios are many times higher than R/C ratios. The upshot is—as pointed out by Barnett and Morse—that physical supply is much less important than rising costs. Yet costs have fallen dramatically over this century for all important energy resources, including coal and petroleum.

Third, any estimates of R/C ratios are a function of technological assumptions. For instance, if nuclear breeder technology is added to fossil fuels, the R/C increases 2000 times. Adding nuclear fusion increases R/C by another 50,000 times.

Fourth, the long-term energy future and the scale of readjustments on the economy and the quality of life depend on the long-run trend in energy prices and on the environmental repercussions of increasing energy consumption and its changing composition. His price forecasts are the shadow prices obtained from applying a programming model aimed at minimizing the discounted costs of meeting a set of final demands:

$$\underset{(i,j,k,l,m)}{\text{Min} \ \Sigma} \ c(i,j,k,l) x(i,j,k,l,m)(1+r)^{-m} \qquad (2.1)$$

where c = production cost, x = activity level, and r = interest rate (assumed 10 percent per annum), i = region (4), j = type and location of resource (17), k = region of demand (4), l = demand category (5), m = 1,2, ... (1970, 1980, 1990, ...), subject to supply and demand constraints.

Although the model is highly aggregative, it is rich in empirical detail. It has an infinite time horizon, but in practice runs for two hundred years. Demand paths are assumed to be linear. The model allocates energy resources efficiently over time and space according to competitive assumptions. It predicts an inevitable transition from a fossil fuel to a nuclear economy. The shadow prices of final energy products increase by 2.2 percent per annum relative to the general price level over the period 1970-2020, by 1.3 percent per annum over the next century, and at a relative zero rate after that. These price increases are of the order of magnitude that can be soaked up by general productivity increases. Of course, some will quibble with the structure of the model, but it is difficult to foresee modifications in its structure overturning the results. The inferences are that "the current energy crisis will blow over eventually," and that "we should not be haunted by the specter of the affluent society grinding to a halt for lack of energy resources."[41]

The environmental impacts may be more difficult to deal with. The programming model included a set of environmental standards, but narrow and not very strict (meeting current standards for sulfur emissions and for nuclear-power production, and including a modest reclamation allowance for surface mining). However, conformity to a specified set of standards to deal with local environmental problems is not insuperable, merely a question of time and cost. The global environmental problems may be more intractable, even though they are more distant. Nordhaus calculated a 43 percent increase in the atmospheric concentration of CO_2 by the year 2030, risky though not fatal, and suggests that a five-hundred-fold increase in waste heat (e.g. 160 years of consumption rising at 4 percent per annum) would be unacceptable.[42] However, environmental risks—though possibly serious—are a very different issue from that of the availability of energy resources.

3 The Theory of Exhaustible Resources

This chapter examines the pure theory of resource extraction, which provides the theoretical underpinnings for the economic analysis of the supply of energy resources through time. It illustrates the recent attention given to natural-resource problems in general, and energy resources in particular, by economic theorists. Although the results are interesting in their own right, and reveal some propositions that shed light on the determinants of resource extraction in the real world, the policy relevance of the analysis is limited by its restrictive assumptions—such as the lack of uncertainty and the existence of futures markets for energy resources.

The Profit-Growth Rule

The central principle of the pure theory of exhaustible resources is the equilibrium condition that marginal profits (or the value of the resource net of extraction costs) must increase over time at a rate equal to the rate of interest. This problem has been analyzed by a host of analysts from Gray and Hotelling to Scott, Gordon, Bradley, Nordhaus, Solow, Banks and Schulze.[1]

We begin by assuming that an exhaustible resource is available in a fixed supply V. The rate of use of the resource over the time period t to T, the date of final exhaustion, cannot exceed V so

$$\int_t^T q(t)\,dt \leqslant V \tag{3.1}$$

Exhaustion takes place when cumulative production, the integral of the output-time function, $q(t)$, equals the total stock of the resource. The problem is to maximize the present value of total profits subject to (3.1) as a constraint. This is a problem in the calculus of variations, but the mathematical details need not be reproduced here.[2] To simplify, the model assumes a single homogeneous resource and complete certainty about the future path of prices and the level of reserves V.

The constrained present-value problem is to maximize

$$NPV = \int_t^T \pi[q(t),t]\, e^{-rt}\, dt - \lambda\left[\int_t^T q(t)\,dt - V\right] \tag{3.2}$$

NPV is net present value; π, total profits; t, time; T is the terminal date;[a] $q(t)$, output; r, the (continuously compounding) interest rate, assumed constant; λ, the Lagrangean multiplier; and V, total physical availability. Profit (π) is interpreted as usual as revenues (R) minus production costs (C). Price is p, average costs are c, average profits are ρ; MR ($\partial R/\partial q$) is marginal revenue, MC ($\partial C/\partial q$) are marginal costs, and $M\pi(\partial \pi/\partial q)$ are marginal profits.

The solution requires determining optimal starting and terminal dates and the optimal output pattern over time. Differentiating with respect to the Lagrangean multiplier yields the constraint as another equilibrium condition. The optimal production pattern is determined by the Euler equation of the calculus of variations, and the boundary conditions are that the function (3.2) attains a stationary value at t and T. The solution is

$$M\pi(t) \equiv MR(t) - MC(t) = \lambda e^{rt} \tag{3.3}$$

The conventional second order conditions apply: $\partial^2 \pi/\partial q^2 < 0$ and $\partial^2 R/\partial q^2 < \partial^2 C/\partial q^2$. With pure competition $p \equiv MR$ and with constant costs $c = MC$, so (3.3) reduces to

$$(p - c) = \lambda e^{rt} \tag{3.4}$$

Also, marginal profits equal average profits at the terminal date

$$\rho(T) = \lambda e^{rT} \tag{3.5}$$

Scott[3] suggests it may be helpful to treat λe^{rt} as a user cost—a sacrifice of future revenue for present sales. This allows (3.5) to be reformulated as an equality between MR and the sum of MC of production plus marginal user costs. This is a useful reminder of the fact that exhaustion is a cost.

To derive the profit-growth rule more clearly it is helpful to replace λ by marginal profits at T. From (3.3), $\lambda = M\pi(T)e^{-rT}$. Therefore,

$$M\pi(t) = M\pi(T) e^{r(T-t)} \tag{3.6}$$

Thus, marginal profits will rise at the interest rate r to a maximum at T. Sometimes it is convenient to set the starting date equal to zero:

$$M\pi(t) = M\pi(0) e^{rt} \tag{3.7}$$

The resource owner is indifferent as to whether he receives a marginal profit $M\pi(0)$ now or a profit $M\pi(0)e^{rt}$ after t units of time. The profit-growth rule

---
[a] In the resource exhaustion case T is the date of extinction. It may also signify the switch date to another resource or the optimal terminal production date.

ensures that the present value of future discounted profits is equal to current profits.

At first sight, it may appear contradictory that an equilibrium condition is that profits increase steadily over time, since we might expect free entry to erode profits. In fact, the model deals with a known finite stock, V, in which property rights are already assigned. The use of the concept of profit is a little misleading, since $M\pi(t)$ is really a scarcity price (or royalty) paid to owners of the resource for the amount extracted. This may be clearer if we use a new term, s, instead of $M\pi$, to represent the royalty per unit of output. Thus, (3.7) may now be rewritten as

$$s(t) = s(0) e^{rt} \qquad (3.8)$$

With this reformulation, it becomes clear that free entry is implicitly dealt with since the model contains a zero profit condition which holds throughout the production period.[4] This condition is

$$p(t) - c(t) - s(t) = 0 \qquad (3.9)$$

To sum up, extraction costs are assumed constant over time. Price equals costs plus royalties, and royalties increase exponentially over the period of extraction until the exhaustion date T. At that date price attains the exhaustion price where nothing is produced or demanded and the resource is used up. The exponential time path of prices from the beginning of extraction to the date of exhaustion is optimal because participants in the market have perfect knowledge about the futures market, and royalties—and hence prices—adjust over the extraction period so that the resource is used up just when the price is high enough to choke off demand. However, the rate of increase in royalties and prices has to be equal to the interest rate (assumed constant), so that resource owners are not tempted by higher (or lower) returns elsewhere to alter their rate of extraction from the optimal rate needed to ensure exhaustion at T.

This simple model contains several interesting features that are central to the pure theory of resource exhaustion. First, the assumption of perfect knowledge and certainty about the future is critical. Second, extraction behavior can be understood only in dynamic and interdependent terms, because production and consumption in each period depends on what happens in every other period. Third, the inclusion of the interest rate in the model shows that resource extraction and use is integrated with the investment process in the economy at large. Holding a resource deposit is an alternative to holding other forms of capital and an alternative to consumption.

Parameter changes in this model affect royalties, price, and the extraction date in obvious ways. For example:

Lower Interest Rate

This implies a higher initial price, reducing initial extraction and lengthening the extraction period.

Increase in Demand

The particular impact of this depends upon the nature of the demand shift, but it always tends to shorten the period of exploitation.

Increase in Stock of Deposits

The effect of new discoveries and other sources of expansion in the stock of reserves is to reduce price and royalties at each point of time, thereby stimulating increased consumption, but also—and perhaps more important—lengthening the extraction period.

Changes in Costs

Cost reductions tend to shift extraction in favor of the present and shorten the extraction period. This is because the initial price does not decline by as much as the cost reduction, implying a higher initial royalty. Uniform cost increases (or unit taxes; see p. 43) imply higher initial but lower later prices and lengthen the period of extraction.

As these examples suggest, it is important in the resource exhaustion model to draw a distinction between the *level* and *rate of change* in the royalty, s. In particular, a high royalty encourages extraction, while a rapidly increasing royalty (relative to the rate of interest) defers extraction and the exhaustion date.

The upshot of the profit-growth rule is obvious. If royalties increase at the rate of interest producers are indifferent as to when they extract the resource, since this time path of royalties equalizes the present value of the resource at all times. Thus, maintenance of equilibrium in the stock market allows the rate of extraction to be determined by conditions in the flow market. Production equals demand at the current price, and the market clears. The demand curve links the market price with the rate of extraction. The increasing royalties over time mean a higher market price, and this will eventually reach the level where demand is zero. The flow and stock markets for the resource are so closely interlinked because of perfect knowledge about future markets, as well as the current market, that the last unit of demand in the final extraction period is also the last remaining unit of the resource stock.

If royalties increase at a rate faster than the rate of interest, it pays not to extract the resource, because its value is appreciating at a rate faster than the net return from reinvestment of the proceeds from selling the resource if extracted. Conversely, if royalties rise at a rate less than the rate of interest, producers will extract more in order to reinvest their net proceeds more profitably than by leaving the resource in the ground. Both these situations will have repercussions on production rates, price levels, and the exhaustion date that are nonoptimal. The case where royalties rise too slowly relative to the rate of interest may be especially dangerous from a societal point of view, because it will induce too rapid a rate of extraction and too early depletion.

Nevertheless, the complete certainty assumptions keep the market from going haywire. If these were dropped, market instability in the form of current dumping or speculative withholding of supply might result. For instance, too slow a rate of increase in royalties boosts current extraction which, with a given demand curve, must depress price. In conditions of uncertainty, and in the real world resource markets tend to be vulnerable to shocks and changing expectations, expectations about future price changes will be influenced by current price levels, and in the case outlined pessimism might become worse. Solow[5] argues that the possibility of instability is not as great as this might imply. Provided that resource owners believe that their resources have a value in the long run, they will not let price fluctuations have too much impact on their behavior. Thus, too slow a rate of increase in resource prices will not lead to cut-throat extraction and dumping but to the acceptance of capital losses on resource stocks. This will permit a readjustment of the royalty time path toward its equilibrium rate.

The Impact of Taxes

The two most simple taxes are a uniform tax on each unit extracted and an annuity tax on the present value of the royalty. The first of these taxes is equivalent in its impact to a uniform increase in costs. Its effect on the present value of s can be reduced by shifting extraction into the future. Earlier prices are higher, later prices are lower and the period of extraction is lengthened, thereby postponing resource exhaustion.

The annuity tax means a tax equivalent to a constant percentage of the royalty s. If this fraction is k, the tax changes the present value of the royalty at each point of time to $(1-k)s\,e^{-rt}$. But this leaves the relationship between every pair of royalties unchanged, and does not disturb the equilibrium condition that s rises at a *rate* equal to the interest rate r. The consequence is that in this case taxes have no effect on extraction behavior. They do not affect the time distribution of the use of the stock, nor the date at which extraction begins, nor even the incentive to discovery.[6] In the sense of its influence on optimality and equilibrium conditions, therefore, an annuity tax is completely neutral.

An Environmental Damages Tax

Resource extraction frequently incurs environmental costs; strip mining is an obvious example. Although some types of environmental damage (e.g. air pollution) are related to current output, others—such as the disturbance to scenic and visual amenity—are plausibly related to the cumulative scale of extraction, i.e. the sum of past and present outputs or the depletion in the resource stock. Consider the latter case. Assume that damages, $D(t)$ are an increasing function of the cumulative loss in the stock $[V(0) - V(t)]$, where $D', D'' \geq 0$. The net present value of discounted future marginal damages is

$$\int_t^T e^{-r(v-t)} D'[V(0) - V(t)] \, dv \tag{3.10}$$

To preserve optimal equilibrium a tax, $f(t)$, as a function of time must be imposed equal to (3.10), the present value of future marginal damages. Thus, the equilibrium condition is that

$$s^*(t) = s(t) + f(t) \tag{3.11}$$

where $s^*(t)$ is the price paid to resource owners net of extraction costs but inclusive of the corrective tax for environmental damages due to extraction.

The imposition of a tax affects extraction behavior. As t approaches T, the exhaustion date, $f(t) \to 0$ because $V(T) = [V(0) - V(t)] \to 0$. The optimal tax rate first increases, then declines, over time. Delay raises the tax rate because this is determined by cumulative extraction, but it also reduces the present burden of a given future payment. Since both the damage rate and the discount rate are exponential their net influence varies over time, even in the special case when the two rates are identical. This is because the tax base (the cumulative loss of stock) itself increases over time at a rate determined by the time path of output, $q(t)$. The net effect of the tax, therefore, is to defer extraction until the later phases of the time horizon. It is also possible for $p(t) < c(t) + s^*(t)$ at some period within the time horizon, implying zero production. The conclusion is that the existence of substantial cumulative externalities, if correctly priced, may alter the time path of resource extraction, and in particular may delay resource exhaustion as implied by a conservation policy.

Recycling

The reuse of already extracted materials is a potentially important means of augmenting a limited stock of nonrenewable resources. This can be introduced into the theory of exhaustion by treating waste recovery as competitive with initial extraction. It is sensible to discuss this problem briefly in terms of a simple model, because the possibilities of recycling energy resources are much more limited than in the case of other natural resources (e.g. other minerals).

It is plausible to assume that recovery costs, c_r, are greater than extraction costs, c_x. If j is the scrap value of waste, the choice between extraction or recovery will depend on the relative magnitudes of $(c_r + j)$ and $(c_x + s)$. Given the assumption about relative costs (and assuming that these remain constant) extraction will dominate if $(s - j) > (c_w - c_x) =$ constant > 0. Since s is a scarcity rent, then increases in the stock of waste might be expected to have a depressing effect on s. However, this impact is severely constrained. The stock of waste is a function of past extraction rates, it is subject to a decay rate, and a fraction of the material (in the case of energy resources a very high fraction) is lost in use. On the other hand, if resource exhaustion were predictable, and if future markets for both unextracted resources and scrap existed, the scrap value j could rise at a rate faster than the value of the pure resource, s. However, this would boost the current rate of extraction, because the demand for waste shifts the demand for output vertically by the value of the recoverable waste. But a futures market for waste extending into the future beyond resource extinction does not exist. In its absence, waste materials considered as a long-term substitute for deposits will tend to be dumped and become a common property resource until needed. If resource extinction were certain, it would be beneficial to society for the government to intervene to set up property rights and a futures market for waste materials.

Recycling is included for the sake of completeness in the theory of resource exhaustion rather than its direct relevance to the energy case. Even if 30 percent of a fossil fuel could be reused, after three recyclings the amount of fuel recovered would only be 2.7 percent of the initial supply. The analogy in the energy field closest to recycling is that of breeder reactors for nuclear power, a very special case in which the fraction of material recovered to that consumed is greater than unity!

Decline in Resource Quality

The assumption of homogeneous resources is very unrealistic. Herfindahl[7] examined the case of two grades of resource and showed that the best grade would be used first. The same result is achieved if the decline in resource quality is continuous. Each grade, g, has to be treated as a different commodity, and therefore, there must be a continuum of futures markets over the period of extraction. Analytically, this can be dealt with by assuming that royalties, s, are a function of grade of resource as well as of time. At any point of time differences in the value of resource rights, s, between grades are due to differences in average extraction costs. Each grade considered in isolation behaves in the same way as a single homogeneous resource with the equilibrium condition that s increases at the interest rate r. However, with sequential use, the value of s for the grade currently extracted will increase at a rate lower than r, because of the increasing costs of extraction. Extraction begins at the highest grade available and s drops to zero when production ceases at T; this follows from the assumption of known and perfect futures markets for all grades of the resource. The interesting finding is that because of these futures markets changes in resource quality have no impact on extraction. Increases in extraction costs are reflected in and absorbed by the royalty paid to initial resource owners.

Technological Change

Probably under the influence of the case of petroleum where (certainly in the Persian Gulf) extraction costs are such a minute proportion of the selling price, the traditional theory of exhaustible resources tends to ignore them. In this event, the rate of increase in royalty payments equals the rate of interest in equilibrium, $\dot{s} = r$. A modification is to allow for extraction costs, but to assume that these are constant, $c = \bar{c}$. This alters the equilibrium condition to $\dot{p} = rs/p$, i.e., the rate of increase in price equals the rate of interest multiplied by the share of royalty in price (which itself increases over time). In both instances the market price of the extracted resource rises over time, and this is the standard prediction of the exhaustible resources models. However, this conflicts with the empirical finding that the relative prices of such resources have *declined* over time.[8]

The contradiction is not difficult to explain. In long-run equilibrium price is determined by the average costs of extraction and the royalty or scarcity payment. Since the latter must in equilibrium increase exponentially at the rate of interest, the only way to explain a declining price trend is in terms of declining extraction costs. Thus, the model needs to allow for technological changes in the resource extraction industry. If it is assumed that costs decline over time due to the impact of technological change, the rate of change of price is equal to the sum of the rate of change of costs (negative according to the assumption) and the rate of change of unit royalties. Resource price may, therefore, decline or rise over time. Suppose it is assumed that unit costs decline at a constant percentage rate α; then we may write

$$\dot{p} = -\alpha c/p + rs/p \qquad (3.12)$$

where $(c + s)/p = 1$. The value of royalties, s, increases exponentially at a rate r, while unit extraction costs decline exponentially at a rate $-\alpha$. Market price, p, is the sum of these, and will typically be a U-shaped function of time. The debate about exhaustible resources hinges very much upon where the economy is on this function (assuming a single aggregate resource). Barnett and Morse[9] show that the United States historically has been operating on the downward-sloping section of the market price-time function. The crucial question is: how long can technological change in resource extraction stave off the impact of increasing scarcity on resource prices? Not for long, the pessimists argue, and indeed suggest that the economy has already moved to the rising section of the $p(t)$ curve. (The price-time function referred to here is, of course, the *long-term* price trend. With a perfect futures market for exhaustible resources all fluctuations and long-run changes in demand would be reflected in the value of $s(0)$. In the absence of such a market current prices may fluctuate widely around their trend value, because of changing expectations in a very uncertain world. Accordingly, the recent experience of rapidly rising prices for energy resources, e.g., world oil prices and their impact on the prices of competing fuels, does not necessarily substantiate the pessimists' case.) The pessimist case is strengthened by the facts that the weight s/p must increase progressively relative to the weight c/p, and that the rate of technological change, α, would have to be very impressive to offset r in an era of very high interest rates.

Switching to a Backstop Technology

Prevailing energy technology is based on resources cheap to extract but relatively scarce in the long run because of finite resources. It is probable that the future will eventually be dominated by a technology using superabundant resources but requiring high capital costs (e.g. breeder or fusion reactors or, even better, solar power). This ultimate technology based on plentiful resources is the "backstop technology." It is important to examine how this enters the basic resource exhaustion model.

Nordhaus[10] has outlined a simple model to illustrate this. Consider two methods of producing electricity. One uses a scarce resource, say petroleum, finite in supply (total supplies = V) and produced at constant unit extraction costs $\bar{c}$. The other uses a plentiful and free resource but requires K dollars of capital per unit of output. Assume that r is the rate of interest, and that demand is inelastic, with d units of electricity demanded each year. If the capital costs of the second method are high relative to the extraction costs of the first, the first method will be used first. Assuming that the units of the resource and of electricity are measured so that one unit of the resource produces one unit of electricity, the switch to the second process (the backstop technology) will occur after V/d years.

The price of electricity, p, at the switch date, $\hat{T}$, is given by the cost of the backstop technology,

$$p(\hat{T}) = (r + \delta)K \tag{3.13}$$

where δ is the depreciation rate. The price of electricity by the first process at the switch date is also $p(\hat{T})$. The royalty on the scarce resource along the efficient path from now until $\hat{T}$ is

$$s(t) = p(t) - \bar{c} = [p(\hat{T}) - \bar{c}] e^{-r(\hat{T}-t)} = [(r+\delta)K - \bar{c}] e^{-r(\hat{T}-t)} \quad (3.14)$$

The royalty on the scarce resource is simply the switch price, $p(\hat{T})$, minus extraction costs discounted back to the present. Similarly, the efficient price path is given by

$$p(t) = \bar{c} + [(r+\delta)K - \bar{c}] e^{-r(\hat{T}-t)} \quad (3.15)$$

This model, simple as it is, illustrates some important points. Three important elements determine the current royalty: the cost of the backstop technology, the interest rate, and the switch date. If the price of the backstop technology is low, if the switch date is a long time away, or if the interest rate is high, then $s(t)$ is relatively low. The switch date may have a powerful impact on the royalty. For instance, a doubling in the switch date (due say to a doubling of reserves V in the expression V/d) would imply a reduction in royalties equivalent to $\sqrt{s/p}$. Another feature of the model is that high interest rates keep royalties low at the outset, but once they begin to rise they soon start to dominate price.

Finally, the limitations of this model must be kept in mind. It assumes complete certainty with a perfect futures market for resources. There is considerable evidence that energy-resource owners reap windfall royalties because of the extreme uncertainties in the future energy situation. In the real world, there are several different major energy resources, each of which has a spectrum of varying grades. Also, there are many different uses and demands. In a complex model calculation of the optimal path and the switch dates becomes difficult.[11]

The Social and Private Rates of Discount

The rate of interest plays a crucial role in the theory of exhaustible resources. Indeed, it is the pivot on which action turns. The equilibrium condition is that the value of the resource deposit (which is equal to the present value of future sales after deducting extraction costs) must increase at a rate equal to the rate of interest. It is important, therefore, that the rate of interest is the "right" one. Provided that private resource owners choose to discount their future profits at the same rate at which society wishes to discount its future consumer surpluses, there is no problem. But what if these two discount rates differ? It is frequently argued that the private rate of discount exceeds the social rate. If this is the case, private ownership of resources will lead to too fast a rate of resource exploitation, and perhaps to premature resource exhaustion, from society's point of view. This is particularly plausible if royalties dominate market price. On the other hand, as Scott[12] has shown, it is much less clear when costs are an important element in price, because of the impact of higher interest rates on costs.

The private rate of discount may exceed the social rate, for several reasons. An important one relates to risks. In the absence of perfect insurance markets for risks, the owners of resources bear risks due to price instability, competition from new discoveries, technological change, the introduction of substitutable resources, and so forth. But many of these are not risks to society as a whole, but risks to individuals because of the possibility of transfers within society. Societal risks are very small relative to *average* income, so that if society makes the decisions the risks, such as they are, are widely spread over the population as a whole. This argument, associated with Arrow and Lind,[13] suggests that shifting decisions from the private to the public sector brings about a more favorable distribution of risks. In other words, it supplies a reason why the private may be higher than the social rate of discount—the fact that the former must allow for much higher risks.[b]

It is quite important in this analysis to separate the Siamese twins of risk and uncertainty. Uncertainties, due to the impossibility of foreseeing the future, are common to both governments and private individuals. They have an impact on behavior, and on discount rates, that is different from risks to profit and capital. In the energy resources field uncertainty is considerable. For instance, technological uncertainties associated with the possible substitution of nuclear for fossil fuels exist regardless of whether ownership and decisions are vested in the public or private sector. Similarly, uncertainty due to the unpredictable decisions of foreign suppliers of fuel is external to the consuming country as a whole, and cannot be internalized in its social decisions. There may be differences in the social discount rates chosen by producing and consuming governments, or even among the producing governments themselves. For instance, the decision by Arab producers to cut back oil production in 1973, ostensibly for political reasons, could be rationalized on economic grounds. Several Arab countries (Saudi Arabia, Libya, and some other desert states) have a lower rate of time preference than either the oil companies or the consuming governments (or certain other oil producers, such as Iran). Because the short-run opportunities for domestic investment are limited if evaluated against the capital gain from oil in the ground, and because of a disinclination to hold too much depreciable foreign currency, the most appropriate method of attempting to adjust social and market discount rates in these countries is to conserve their oil reserves.

---

[b]Arrow and Lind suggest that the riskless rate of discount should be interpreted as the social rate. Hirshleifer and Shapiro, on the other hand, suggest that the social rate varies according to the *risk class* into which the social investment opportunity falls. This can be measured via the private rate of return to investments in that particular risk class (J.H. Hirshleifer and D.L. Shapiro, "The Treatment of Risk and Uncertainty," *Quarterly Journal of Economics*, 77 (1963), pp. 95-111).

In the light of recent events, it is probable that governments in general—and the US government in particular—will attempt to minimize uncertainty more than in different sets of circumstances (e.g. the less uncertain days of the 1960s). One possible outcome is that the government will become too defensive, and choose a rate of discount that is too low. The strategy of conserving resources and boosting prices (discouraging consumption and promoting the search for new discoveries as by-products), merely as a protection against uncertainty, may impose disproportionately high costs on present consumers, to the benefit of those in the future. How individual resource owners might react is more difficult to predict. In a model assuming certainty, resource prices rising faster than discount rates induce conservation and delay production, but in an uncertain world resource owners, if they believe that current prices are artificially high because of interferences with the market, and fear that the bottom might drop out of resource markets in the not-too-distant future, may attempt to cash in on the current situation by extracting as fast as they can. The plausibility of these scenarios is not important. What is important is that the existence of uncertainty can result in deviations of both the social and private rates of discount from the social rate that would prevail in a certain world, that the social rate in conditions of uncertainty might be too low rather than too high, and that higher costs imposed on the present generation deserve attention as much as looking after the future generation.

Another problem, discussed in detail by Baumol,[14] is the distorting effects of taxes on capital income, such as the corporate income tax. Government policy appears to treat the long-term rate of interest on government bonds as an appropriate discount rate for public investment decisions. Baumol argues that if individuals are prepared to purchase long-term government bonds at an interest rate of r percent, it is not unreasonable to use this rate as a proxy for their rate of time preference. Unfortunately, the existence of capital taxes destroys the possible equality between the individual time preference rate and the private rate of return to investment.

This can be shown with a simple model. Assume full employment, zero risks, corporations financed solely by equities, a uniform corporate tax rate of 50 percent, a single rate of interest, r, at which the government borrows money, and no taxes on individuals or noncorporate producers. The inputs used by the government in its investments have their sole alternative use in corporate investment. The opportunity cost of using these resources is, therefore, their private sector rate of return. With a 50 percent corporate tax rate the equilibrium rate of return in the private sector would have to be $2r$. If the social discount rate is related to the before-tax rate of return in the private sector, it will induce too fast a rate of resource depletion from society's point of view. Some people might use this argument to support the favorable tax treatment granted to the petroleum industry rather than the high-risks justification frequently suggested.

The existence of such taxes places us in a second-best world, where the economist is somewhat vulnerable. The appropriate solution, from the viewpoint of efficient allocation, might be to repeal the corporation tax, but this proposal is hardly likely to win a prize for practicality. If natural resources are judged to be a special case, one solution might be to grant the industries involved favorable discrimination in tax treatment. To some extent, this already exists. How far it should be taken, and what are its implications for allocative efficiency, are anybody's guess. A simpler and safer, though more ideological, answer might be to vest ownership and extraction of natural resources in general, and energy resources in particular, in the hands of the government. Even here Solow's warning must be kept in mind:

it is far from clear that the political process can be relied on to be more future-oriented than your average corporation. The conventional pay-out period for business is of the same order of magnitude as the time to the next election, and transferring a given individual from the industrial to the government bureaucracy does not transform him into a guardian of the far future's interests.[15]

A quite different argument for the belief that the market might exhaust resources too quickly is the view that private time preference is an improper basis for intertemporal decisions. Frank Ramsey[16] in a classic paper argued that future utilities should not be discounted by society. There is no justification for discriminating between generations, and so we ought to behave as if the social rate of time preference were zero.[c] This view relies far more on ethics than on economics. But it is undeniable that choosing the appropriate social rate of time preference is a critical, one might even say a life-or-death, question in an economy with a fixed pool of exhaustible resources. Dasgupta and Heal[17] and Solow[18] have shown that the optimal path with a positive discount rate might lead to consumption per head going asymptotically to zero (due to resource exhaustion), whereas a zero discount rate might result in perpetually rising consumption per head. Solow suggests that the choice of the social discount rate is a decision about intergenerational equity. He adopts the "highest-constant-consumption" criterion, according to which the planning problem becomes how

[c]Herfindahl, among others, has argued that this is unacceptable because it induces a transfer of consumption from people with lower incomes to those with higher incomes, assuming the continuation of growth (O.C. Herfindahl, "Goals and Standards of Performance for the Conservation of Minerals," *Natural Resources Journal*, 3 [1963], pp. 78-97). A zero discount rate demands of the present generation that it must be worse off to make future generations even richer. However, a zero utility discount rate is compatible with a positive consumption rate of discount if the marginal utility of income diminishes with increasing income and if per capita incomes are expected to be higher in the future. Also, the picture changes more than a little if the prediction of wealthier descendants is faced with the counterprediction of no descendants at all! This latter scenario may sound hysterical, but the equation of resource exhaustion with race extinction is a familiar identity in the Doomsday literature.

to maximize consumption per head with the requirement that it should be constant over time, thereby preserving equality between generations subject to constraints, including finite resources. Industrial societies have become so used to rising expectations that this criterion may be inequitable to future generations. A satisfactory living standard of 100 years ago would be regarded as intolerable poverty today. On the other hand, the chastening experience of the 1970s hitherto may suggest that it is about time we began educating ourselves out of rising expectations.

The problems associated with resource exhaustion may be deferred by technology, either by technical progress—especially of the natural-resource-saving variety—or via the substitution of capital and labor for exhaustible resources. Solow examines the zero technical progress case, and shows that if the elasticity of substitution between exhaustible resources and other inputs is not less than unity, and if the elasticity of output with respect to capital exceeds output elasticity with respect to natural resources, then a positive constant consumption per capita can be maintained indefinitely with a constant population. Since this consumption level is an increasing function of the initial capital stock, resource drag can be avoided if the initial capital stock is large enough. If the elasticity conditions are not met, however, the long-run maximized consumption level is zero. The virtue of these new theoretical writings is that they focus attention on the key variables in the issues connected with exhaustible resources: the appropriate discount rate, the technical progress potential, and the scope for factor substitution.

Before leaving this analysis, it is worth mentioning the argument of Gordon[19] that the stationary state, or "bliss," associated with the Ramsey model is incompatible with exhaustible resources. The point is that there would be no resource extraction in a permanent stationary state with a zero rate of interest. Why? Return for a moment to the equilibrium condition that the value of resources must grow at a rate equal to the rate of interest. In a zero interest world it is profitable to wait forever before mining, because resources cannot have negative values. Thus, there is a clear contradiction.

On the one hand, the mining of any material which can yield services that exceed the cost of the services required to produce it is an investment with a positive yield; thus elimination of these possibilities is required before capital glut is attained. On the other hand, it simply is not possible to have mineral production in a permanent stationary state with zero interest.[20]

The practical solution to this theoretical puzzle is presumably that resource use in the stationary state would take the form of continuous recycling of existing materials. This might be all right for natural resources in general, but it does not apply to fossil fuels, which are largely destroyed in use. The answer in this case would have to be something like solar energy.

Externality arguments may also be used to support the hypothesis that the private exceeds the social rate of discount. For instance, if utility functions are interdependent, comparable, and additive both within and between generations, it is possible to outline cases where individuals would be willing to save more if they knew that others were saving more also. Thus, a social saving program—financed via a uniform tax, for example—would drive down the discount rate lower than when all saving is carried out independently by individuals.[21]

This may be shown with an example. Suppose individual i regards $1.00 of future consumption for society as equivalent to 8 cents of his present consumption. Further assume that, because investment is productive, $1.00 of current investment provides $2.00 of future consumption. In this case, i will not save the dollar, because the psychic gain in return for the dollar is only 16 cents (2 x 8 cents). If the decision were being made socially, however, the outcome may be different. Now individual i knows that his sacrifice will be matched by sacrifices imposed on everybody else. Thus the psychic gain is now $2n(.08)$. However, individual i also incurs a psychic cost as a result of consumption sacrificed by his contemporaries. Assume that the trade-off is $1.00 of a contemporary's consumption for 15 cents of i's consumption. Assuming that everyone has the same preferences, the marginal time preference for each person is the ratio of psychic gain to him from future consumption to psychic loss from sacrificed present consumption. In the example this ratio is: $2n(.08)/1 + (n-1)(.15)$. This ratio exceeds unity if $n > 85$.

This is an interesting theoretical argument, but it requires some strong assumptions, particularly about intergenerational transfers and the degree of interdependence among utility functions. Moreover, the model works only if future consumption from a given amount of current investment is valued more highly than the present foregone consumption needed to finance that investment. In the example above, if the trade-off between future consumption and an individual's present consumption had been $1.00 to 7 cents instead of 8 cents, the saving would not have been justified even if made collectively.

Part II:
Energy Resources

4 Oil

Trends

Recent trends in the United States petroleum industry are illustrated in Table 4-1. The reserves position has remained more or less stationary since the 1950s, apart from a slight boost in 1969-70 due to the discovery of oil in Alaska. Production of crude increased steadily up to 1970, but has stagnated since that date; offshore production, which also increased steadily in the 1960s, has remained stable since 1970. However, this lack of growth should not necessarily be extrapolated into the future, since Alaskan oil and new offshore fields should start production in the next two or three years.[a] The crux of the "energy crisis" lies in the data of the last column of Table 4-1—the growth of imports. These have soared since 1970, and in combination with stagnating home supply they imply that all the increase in petroleum demand has been met from foreign sources. If oil has strategic importance economically and politically, then an import ratio of close to 40 percent may be too high for a country with the ambitions and obligations of the United States. Moreover, the physical quantities underestimate the impact of imported oil. The sharp rise in prices since 1973 has resulted, despite the return of oil revenues in the form of additional demand for US exports, in severe balance-of-payments strain. The cost of oil imports rose from $3.9 billion in 1972, the last "normal" year, to $24 billion in 1974. These trends explain why US policy-makers are so keen to moderate the growth in petroleum demand and to stimulate domestic oil production.

The history of the United States domestic petroleum industry is a history of output restriction and limitations on competition. To ask this industry to expand output as rapidly as possible is to ask it to change its traditional habits of many decades. The difficulty of achieving this may be clearer if some of the historical methods of restricting competition are briefly examined. These factors should be considered alongside the more obvious shorter-run constraints such as the discovery of new fields, the lead-times for opening up new wells, and the shortages of drilling equipment—especially for the smaller, independent oil companies.

---

[a]A possible bottleneck to progress in the exploitation of offshore oil was eliminated in March 1975 by the Supreme Court's decision that the federal government, not the states, has jurisdiction over offshore fields.

Table 4-1
United States, Oil Reserves, Production and Imports, 1950-73

	Proved Reserves (billion bbls)	Crude Oil Production (MBD)			Imports (MBD) of Crude Oil and Petroleum Products
		Total	Offshore	Offshore as % of Total	
1950	25.3	5.41			0.85
1955	30.0	6.81			1.25
1960	31.6	7.05	0.32	4.5	1.82
1965	31.4	7.81	0.67	8.5	2.46
1966	31.5	8.30	0.82	9.9	2.57
1967	31.4	8.81	1.01	11.4	2.54
1968	30.7	9.12	1.29	14.2	2.85
1969	29.6	9.24	1.44	15.6	3.16
1970	39.0	9.64	1.58	16.4	3.41
1971	38.1	9.46	1.68	17.8	3.92
1972	36.3	9.48	1.56	16.4	4.75
1973	40.0	9.21	1.60	17.4	6.16

Sources: American Gas Association, American Petroleum Institute and Canadian Petroleum Association, *Reserves of Crude Oil, Natural Gas Liquids and Natural Gas in the United States and Canada and United States Productive Capacity* (May 1973).
U.S. Geological Survey, *Outer Continental Shelf Statistics* (June 1973).
U.S. Bureau of Mines, *Minerals Yearbook*.

Rule of Capture

"Rule of capture" is a principle established by law which in the absence of other legislation and controls to mitigate its effects militates against conservation of petroleum resources. Assuming that an underground oil reservoir crosses over surface property lines, any landowner or lessee may withdraw oil from the reservoir freely, regardless of its impact on his neighbors. If several producers share a common pool it is in the interest of each to extract as much oil as possible as quickly as possible. In this case marginal user costs are negative, because each barrel produced reduces the potential loss of output to neighboring producers. Where it applies, this negative marginal user cost tends to outweigh the conservationist effect of positive marginal user costs—the fact that present output sacrifices higher value future sales. Thus, rule of capture distorts the optimal path of resource extraction by discriminating heavily in favor of present production. It also gives rise to physical waste, since too fast a rate of extraction frequently reduces the total amount of oil recoverable from the pool.

Fortunately, many oil-producing states have legislation that prevents rule of capture from having disastrous effects. Well-spacing rules, for instance, reduce the probability of heavy competition among producers in the exploitation of a single pool. However, well-spacing controls tend to be too arbitrary, since they impose a minimum distance between wells, whereas the appropriate minimum distance (e.g. to ensure that each well produces at its MER—the maximum efficient rate) will vary according to the specifics of the individual case. Prorationing devices will also offset rule-of-capture distortions, because they determine the production quotas for each producer. The most effective protection against rule of capture, however, is field unitization, which in effect creates single management of each field. This solves the rule-of-capture problem, since its necessary preconditions are a reservoir spreading over multiple surface rights and several producers.

Prorationing

"Prorationing" is the term used in the oil industry to describe the old-fashioned cartel practice of restrictive production quotas. Two basic types have been used. Market demand prorationing shares estimated demand among existing fields so as to "preserve equity." MER prorationing allocates production according to the MER of each well. Of the two, the latter is much less restrictive, because it allows low-cost wells to produce more than under market demand prorationing. However, it is still not preferable to allowing output to be determined by competitive forces. MER production avoids geological waste, but the optimum geological rate is not necessarily the economic optimum.[1] For instance, where rule of capture operated, producers would all be encouraged to produce at the MER, and this could lead in some circumstances to overproduction and too fast a rate of resource depletion.

Nevertheless, market demand prorationing is much more inefficient. Its primary effect is to shift output from low-cost to high-cost producers, involving a leftward shift in the industry's marginal operating cost curve, and a higher oil price. The efficient producers are restricted to extraction levels below the profit-maximizing output. Industry output is virtually determined by decree, subject to the constraint that each well produces an output consistent with marginal revenues being equal to or greater than marginal prime cost.

In current market conditions, prorationing is dormant. However, if the domestic oil industry received heavy insulation against competitive imports, there might be circumstances under which it might be profitable for such restrictive practices to be revived.

Field Unitization

Many of the regulations that litter the US petroleum extraction industry were developed in a piecemeal fashion, primarily to deal with the inefficiencies and waste associated with several producers simultaneously trying to exploit one field. It has been argued by many observers (e.g. McDonald[2]) that there would be little justification for most restrictions if each field operated as a single enterprise under a unified management. If this could be achieved, either on a voluntary or a compulsory basis, it might be possible to dismantle all the other regulations and interferences with competition and to allow each field to compete with the others. Such amalgamations—called "unitization"—would allow each surface owner of a field to share equitably in a production company that would extract the field as a whole. Its advantages include avoidance of the rule of capture and allowing the low-cost wells in a field to be used more intensively. Given certain assumptions about the equation of private and social rates of discount and of private and social costs of production, an industry of unitized fields would tend to choose the optimally distributed rates of output that—in the absence of uncertainty about future prices—would both maximize profits for the individual operator and be consistent with a societal optimum. Although this compatibility between private and public interests requires some unrealistic assumptions, unitization offers substantial benefits in comparison with the alternative system of overlapping, arbitrary, and sometimes conflicting regulations.

There are obstacles to effective unitization. Several states have compulsory unitization laws, but these prescribe stringent preconditions such as a very large majority of the property owners affected being in favor. Voluntary unitization may be opposed for a variety of reasons, from desire for independence to clear self-interest (e.g. the structural advantage of wells furthest away from the driving fluid under water or gas-cap drive). The leading producing states have been among the most reluctant to consider compulsory unitization. Although the share of national petroleum output produced from unitized fields has increased since the 1940s, it still remains a minority of the total.

Moreover, encouragement of unitization runs a risk common to many types of merger promotion activity—namely, strengthening monopoly power. In the petroleum industry, this risk is greatest in the case of the large integrated refiners, who are frequently the biggest producers in each field.

Taxes and Petroleum Extraction

Taxes affect the oil industry in several ways: corporate and personal income tax, property, severance and excise taxes, and import duties. However, the petroleum industry has been treated uniquely in three respects: percentage depletion, the expensing of intangibles, and privileged treatment under the foreign tax credit scheme. Any discussion of favorable treatment should focus on these aspects of the taxation of oil income.

Taxable business income is gross receipts minus legitimate costs. Among these costs capital consumption is recognized as a permissible deduction. In manufacturing capital consumption (the depreciation allowance) is usually measured as a percentage of the original cost of acquiring the capital assets. In mineral extraction the producer has had the alternative of subtracting a predetermined percentage of income from production that is not related to original cost. In the case of oil and natural gas this figure (percentage depletion) was fixed at 22 percent of gross income, subject to the constraint that it could not exceed 50 percent of net income (before depletion). The rationale for this procedure—determined historically—was that there is no clear relationship between discovery costs and the value of deposits discovered. The effect of this provision was that a successful resource owner could recover several times his original capital costs over the lifetime of his resource.

Expensing allows oil and gas operators to write off as current expenses intangible drilling and development costs of wells, i.e., all costs other than tangible costs with a secondhand value, such as pipes, pumps, and tanks.

Many arguments have been used to justify these privileges: the need to stimulate exploration; the protection of national defense interests; compensation for the extreme riskiness of crude oil production; the fact that oil is, in effect, a capital asset—indeed, a wasting capital asset—and the net revenue of the industry should not be treated solely as taxable income; and—a more sophisticated argument—a uniform corporate income tax is *non*-neutral if it is shifted forward as higher prices, and the tax provisions for the oil industry help to reneutralize the effects of the corporate income tax.

The counterarguments are also numerous: low tax rates and high rates of return go uneasily together when tax rates are higher and rates of return lower in other industries; depletion was "primarily an ad valorem subsidy to mineral rights owners who have no alternative use for their resource"[3]; the supply of oil lands is inelastic; and, perhaps most important of all, a lower effective tax rate attracts capital into the industry more than in a free market, drawing it from marginally more efficient outlets and reducing GNP. Another set of objections challenges the specifics rather than the principle. Expensing may be preferable to percentage depletion because it is a tax benefit at the exploration-and-development phase, rather than when resources have been proved and extraction begun; the 22 percent depletion may have been inappropriate,[b] and may have led to interfuel misallocations (the depletion rates for other fuels were lower); and the national security objective would be better served if the tax privileges applied only to domestically produced oil, rather than to the operations of US firms as a whole.

These issues are very complex. Much depends on the neutrality argument. If entry is free, higher rates of return in oil extraction are compatible with efficient resource allocation if the industry is subject to special risks. Where industries

---

[b]The depletion rate was 27.5 percent before 1969. The allowance was finally abolished for large producers and phased down for the smaller independent producers in the 1975 tax bill.

differ in capital intensity and rates of return (e.g. due to risk differentials) the shifting forward of a uniform tax alters relative prices and the allocation of demand. The tax burden of the industry should not be considered as a whole; that is, property and severance taxes should also be taken into account. However, even if these arguments were sound, the inference is not special treatment for oil, but a more neutral tax system for the economy as a whole.

The stimulus-to-exploration argument is unconvincing. The supply elasticity of oil lands is difficult to establish. The high levels of investment in oil exploration are coincidental with, rather than induced by, lower taxes, since exploration rates were high before preferential tax treatment existed, and are high in countries without such benefits. The critical question of whether *additional* exploration was induced cannot be answered. In any event, the percentage depletion provision was an inefficient form of subsidy to exploration, since in effect it subsidized extraction. It is arguable that it altered the time path of extraction, shifting production in favor of the present. This hastens resource depletion, and runs counter to societal resource-conservation strategies.

An even greater advantage than the favorable domestic tax treatment for the integrated multinational oil companies has been the foreign tax credit. Under this arrangement US companies may credit most foreign income taxes against the United States income tax liability on foreign income, rather than claim such foreign taxes paid as a deduction from foreign taxable income, either on a per-country or overall basis. Although this applies to all companies, the oil companies have enjoyed the special benefit of being allowed to credit taxes which are, in effect, royalties to the oil-producing countries. This arrangement was started in 1950 with the Saudi Arabian government. If these levies had been appropriated as royalties, the payments could only have been used as an income tax deduction rather than as a tax credit against the US income tax liability of the oil companies. The result has been that the US multinational oil companies have had a lower tax liability (foreign plus US) than if this tax advantage had not been allowed.[4] However, these advantages were reduced in the 1975 tax bill.

The Power of the Oil Industry

In terms of size, influence, and potential economic and political power, the oil industry is probably the most important industry in the United States. Many believe that it has frequently operated against the public interest, but such a hypothesis is difficult, if not impossible, to test. The supporting evidence is circumstantial, and there is some evidence on the other side.

Although oil companies account for four out of ten of the nation's largest industrial corporations, and the concentration ratio (share of total output accounted for by the four biggest firms) increased from 19 to 31 percent between 1955 and 1970,[5] this remained lower than in many other industries—both within and outside the energy sector. The oil industry has diversified extensively in the 1960s into other energy fields—gas, coal, shale oil, and uranium—but the economic forces in support of such a development are strong. The giants in the industry work closely together in pipeline transportation, product exchange agreements, refining agreements, and joint exploration ventures, but if this provides the opportunity for collusion it would be wrong to infer that collusion therefore exists. Barriers to entry mainly take the form of high capital costs for bidding for leases, exploration, and drilling. The price and profit record of the industry before 1973 was reasonable. Relative prices fell in the 1950s and 1960s, and rates of return were comparable to the average for all industries. If prices (on new oil) and profits have soared since 1973, this is a by-product of the actions of OPEC. Other measures taken by OPEC, namely nationalization of production and refining operations, may—despite handsome financial recompensation—be inimical to the oil companies in the long run. Moreover, even if oil is the pacemaker, there is a degree of interfuel competition. It would be hard to challenge the argument that the industry is workably competitive.

Perhaps a more serious charge is that the oil industry wields disproportionate political power, assisted by the pluralist political structure that offers opportunities for effective lobbying and the exercise of vested interests. In terms of industry size, jobs at stake, geographical dispersion, powerful trade associations, and money for advertising, lobbying, and political campaigns, the energy industry in general and the oil sector in particular has more opportunity than any other for making its influence felt. Oppenheimer[6] has shown how the oil industry makes itself heard in Congress, and historically there are many examples of legislation and executive decisions favorable to oil interests. They include the prorationing and other restrictive devices virtually sanctioned by law, the protection given to domestic producers by the import quota program of the 1960s, the discriminatory tax advantages, the postwar failure to take effective antitrust action, and leasing policies that favor the large companies and have resulted in a higher concentration of reserves ownership than in production. Whether or not this influence is waning is difficult to establish. The action of Congress on percentage depletion in 1975 might suggest that perhaps it is, but that may reflect its political complexion and the fact that fat profits have weakened the resistance of the industry itself. Evidence in the other direction is the administration's change of mind about taxing "windfall" profits that might result from higher gasoline and heating oil prices associated with an energy conservation and import control program. Rising energy prices have probably awakened the public to greater concern and involvement, but until this is translated into effective impacts on decision-making processes the oil industry will remain a power in the land.

The Oil Import Question[c]

History

The issue of oil imports—their role in US domestic consumption and their impact on domestic oil production—has long been of great concern, and has attracted much official attention. However, in the light of effectiveness of the OPEC monopoly cartel during and after the Arab oil embargo of 1973, the past debates have an air of unreality. These debates did not concentrate on issues of principle, but on the practical effects of alternative protection strategies, and evaluation of these effects depended upon the assumptions about price levels. Even as late as 1970 the comprehensive study by the Cabinet Task Force on Oil Import Control (CTFOIC) looked ahead to 1980 in terms of price forecasts that were, even taking the extreme of the upper range, well below one-half of the current price levels. Any calculations made by that or any earlier body are worthless, because estimates of elasticity of demand, growth in energy consumption, elasticity of supply of domestic fuel production, and other parameters needed to evaluate energy problems and policy refer to a set of conditions outside their imagination, not to mention their experience.

Nevertheless, leaving aside the irrelevancy of the older calculations and nitpicking about estimates of the costs to consumers of alternative levels and types of import control, a brief discussion of some of the principles involved provides a useful background for analysis of contemporary issues such as Project Independence. Recent experience, however, has shifted the ground of the debate. Although a minority opinion, prior to the 1970s the free-trade argument could be discussed seriously. The cost differential between Persian Gulf oil (even after inclusion of transport costs) and US domestic oil was wide enough for the low-cost-energy-supply argument to have some appeal. Comparative *costs* remain wider than ever, but the evolution of a—for the moment—highly disciplined selling cartel and the invasion of political motives into supply decisions make the gains-from-trade thesis attractive only to deeply buried ostriches. For the foreseeable future the real issues are how much protection, by what methods, and at what cost. The options range from continuing reliance on "safe" supplies from abroad reinforced by insurance devices such as oil storage banks to total independence of foreign supplies by adopting an expensive self-sufficiency strategy. Although there are many possibilities within this spectrum, the range of feasible and politically acceptable alternatives has narrowed considerably compared with a few years ago.

With the exception of the special case of 1920-22, the United States was a net exporter of crude oil and refined products up to 1948. However, the rapid growth in energy consumption was too fast for the domestic supply to cope at

[c]The discussion in this section presents a background to the appraisal of the international oil situation (Chapter 8) and national energy policy (Chapter 12).

prevailing prices, and it became economically expedient to allow oil imports from cheap and abundant sources to drift upward. In the first year of net importing, the excess of imports over exports was only 50 million barrels. By 1970 oil imports had reached 3.4 MBD (million barrels per day), and by 1974 6.3 MBD. Forecast imports are a function of forecast prices. If the price was $11 per barrel in 1985, oil imports might fall to 3.3 MBD; if the price fell from current levels to $7 per barrel in 1985 imports in the "no response" case might rise to 12.2 MBD. An "accelerated supply" strategy might cut imports to 8.5 MBD at a $7 world price; if reinforced by a conservation strategy this level might be reduced even further to 5.6 MBD (Table 1-6, p. 19). Thus, even in the most optimistic cases post-1973 experience is not expected to wipe out import dependence altogether (unless world prices are very high and accelerated supply and conservation measures are very tough).

A tariff on oil imports had been in existence since 1932, and compulsory import quotas were briefly introduced in 1933. Both the industry and the government became worried about the trends in the early 1950s, if for different reasons. The pace of discoveries, particularly in the Persian Gulf region, and the impact of economies of scale in tanker transportation on transport costs rightly increased the fears of domestic oil producers about foreign competition, while the government was more concerned about the threat to national security if the ratio of imports to total demand became too high. A request to oil importers to keep imports in balance with domestic production in 1955 was not successful, and in any event the Suez crisis of 1956 intervened. In mid-1957 a detailed voluntary quota system was introduced, but soon failed because of noncooperation by some importers and diversion into refined and semirefined imports. Thus, a mandatory import quota system was adopted in March 1959. With the exception of the West Coast, imports other than residual oil were fixed at about 9 percent of domestic consumption (reformulated in 1962 as 12.2 percent of domestic production of crude and natural gas liquids). For the West Coast and for residual oil, imports would be permitted to fill the gap between demand and domestic supply (in April 1966 the restrictions on residual fuel oil imports were in effect lifted). This system—with minor amendments—worked reasonably well in constraining the growth of imports, though more effectively for crude than for residual.

However, events associated with the Arab-Israeli war of 1967—such as the temporary stoppage in production by Arab producers, the partial embargo, the shut-down of the Trans-Arabian pipeline and the Suez Canal—brought home the gravity of the national security problem, and in 1969 the CTFOIC was set up to reexamine the question. The CTFOIC argued that security of supply required "an absolute maximum" ceiling for *Eastern Hemisphere* imports equal to 10 percent of domestic demand.[7] For the first time, differential risks were accepted as being attached to various sources of supply. After a transition period, it was recommended that Canadian oil be allowed to be imported freely (provided that

US-Canadian import policies were harmonized), and that Venezuelan oil be given some preference (initially a quota, later a fixed tariff per barrel).

A majority of the CTFOIC recommended replacing the quota by a variable, differential tariff, and a slightly smaller majority suggested a tariff system that would reduce the prevailing US crude price by 30 cents a barrel. Perhaps because of the lack of unanimity, no action was taken on the CTFOIC's recommendations: "the Administration's unwillingness to solve the problem of dependence in 1970 ... put off the hard decision of supply reliability for four crucial years."[8] In fact, the CTFOIC proposal of a variable tariff receives considerable support today, perhaps reinforced by a storage scheme to insure against intermittent and unpredictable interruptions in foreign supplies.

In April 1973 the Mandatory Oil Import Program was abolished along with all tariffs on oil and refined products, and replaced by a license fee quota system. Holders of import licenses were allowed to import oil up to the quota allocations prevailing in 1973. Imports in excess of this level were permitted in return for a small fee paid by the importer. This change was undoubtedly influenced by the changing supply-demand position in the United States. The expressed objectives of the reform were to help meet energy needs by encouraging oil imports at lowest cost; to encourage refinery construction in the United States by imposing a higher fee on refined products than on crude oil (also, crude oil could be imported without fees for five years in amounts up to three-quarters of new refining capacity); to maintain short-run and long-run flexibility by being able to adjust the fee in response to changing conditions; and to contribute to national security by keeping imports under constraint.

Tariffs vs. Quotas

Between 1959 and 1973 a quota system was the main form of protection from imports to the US domestic oil industry. However, the CTFOIC came out in favor of a tariff, though this was not adopted at the time. But the oil import crisis of 1973-74 led to a suspension of the quota scheme, and much of the recent debate between the administration and Congress has hinged on the relative effectiveness of tariffs compared to quotas. The choice of protection instrument may be important, especially if world oil price levels were to fall (as predicted by some observers, such as Adelman[9]). It is useful, therefore, to rehearse some of the main arguments influencing the choice between the two main devices.

As is well known in the international trade theory literature, a tariff and an import quota can be equivalent, assuming no uncertainty. The main difference is that tariff revenues accrue to the government, whereas the assumption with quota benefits is that they accrue to the industry. In conditions of uncertainty, however, the choice between a tariff and a quota can make a difference, because supply and demand responses may differ in the two cases. The CTFOIC proposal, in fact, was for a tariff-quota system rather than a pure tariff. Eastern Hemisphere oil would be subject to a variable tariff and a maximum quantitative limit. After the transition period, imports from the Western Hemisphere would be subject to a fixed tariff but no quota restrictions. The CTFOIC's justifications for tariff protection were that it would promote more efficiency in the use of resources; it evades the problem of having to allocate import rights; and it makes it more difficult for nonfederal agencies to control domestic output and prices. In other words, a tariff at least subjects domestic producers to the stimulus of competition, however blunted, and it is easier to administer.

The counterarguments stress that it depends upon what the objectives of protection are. In the security case it might be argued that the primary aim is to restrict imports to a maximum share of total demand, and this can be achieved more effectively by a quota system at variable costs (competitive bids for limited import rights) than by a tariff, variable imports at a fixed cost. The alternative of a variable tariff requires the government to be able to forecast demand and domestic supply in advance in order to determine the appropriate tariff, unless it is automatically pegged to a particular domestic price.

The emphasis in current import control discussions has shifted from the issue of efficiency vs. protection to the needs to reduce the import bill, stimulate domestic production, promote energy independence, and encourage conservation. In this case the choice between a tariff and a quota largely hinges on value judgments about the effectiveness of price signals and the importance of equity considerations.

Security

The security aspect of petroleum supplies and hence import dependence is no longer a debatable issue. The events of 1973 saw to that. Perhaps the emphasis in the security argument has shifted. The military aspect (the fact that petroleum is given equivalent status to arms and ammunition) is relatively less important—or perhaps merely more distant. Of course, the argument that interruption to oil supplies for military use is less pressing because of the existence of the nuclear deterrent is dangerous and unconvincing. The diplomatic security thesis—that dependence on foreign energy sources inhibits the execution of an independent foreign policy—is nebulous, though clearly important to the State Department under Kissinger. The strongest security threat is to the civilian economy, because of its heavy dependence on petroleum and natural gas, presently about three-quarters of total energy consumption—and virtually 100 percent dependence in the transportation sector. This threat does not in itself imply a need for self-sufficiency. The long-run problem is to hold down imports to a share of total demand that does not put the economy at risk if supplies cease to be available on a more or less permanent basis. The short-term problem is to avoid or to minimize the readjustments to the economy necessitated by temporary interruptions in supply. This can be dealt with in several, by no means mutually exclusive, ways: a minimal import share; discrimination among suppliers according to estimates of their reliability and friendliness[d]; temporary controls to combat oil shortage; and emergency storage banks.

The security question dwarfs all other arguments for protection. For instance, the need to maintain the size of the domestic petroleum industry is convincing only as a subset of the security problem. Conversely, the opposing argument that allowing imports to expand freely conserves domestic resources for the future when resource constraints may be tighter has too high a security cost in view of the long lead-times needed to get petroleum production under way.

The question of the cost of an adequate security protection strategy cannot be examined in detail in the absence of using specific data on prices, costs, demand and substitute fuels. The experience of past attempts in this direction, and their tendency to be overtaken by events, is no stimulus to a new attempt here. It is much safer to refer to some of the factors that must be taken into account in any evaluation.

First, costs can be estimated only for a very specific type of import control policy. For example, the costs of a tariff relative to those of a quota impose different burdens on consumers, even in the case where a tariff is calculated so as to generate the equivalent flow of imports to a stated quota. Also, the introduction of a storage policy allows a lower tariff or a higher quota for a given level of security.

Second, the case for maximizing imports subject to attainment of a desired level of security is valid only if world prices for oil are substantially lower than the US price. Unless world prices drop markedly from their current levels, the low-cost argument against import control falls to the ground. The substantive point, independent of current conditions, is that the costs of protection depend on the US-world-price spread, and on the necessary condition that the US price should be higher.

Third, a distinction must be drawn between costs to consumers of oil products per se and resource costs. If US oil prices are higher because of import controls, many of the "costs" are merely transfer payments from consumers to producers, resource owners, and the government. Resource costs arise only to the extent that higher prices in the petroleum industry attract resources from other sectors where they would be more profitably employed in the absence of

---

[d]In these hard times few suppliers could be described as friendly. Venezuela is a hard-liner on price, and prominent in OPEC. In November 1974 Canada, realizing that earlier estimates of reserves had been too optimistic and facing the possibility of becoming a net oil importer after 1982, announced a plan to cut oil exports to the US from 0.9 MBD to 0.65 MBD by July 1975, with more cuts to follow. Withholding Albertan oil from export would also make completion of Canada's pipeline to the East viable, enabling her to economize on imports from Venezuela and the Middle East.

the protection, and on the consumption side, because high fuel prices will tend to induce substitution of expenditures on fuel for other types of expenditure or for savings.

Fourth, interdependence with other fuels—especially natural gas—has to be considered. To the extent that oil and natural gas are joint products (the CTFOIC estimated that 25-30 percent of gas additions were attributable to oil exploration), repercussions of import control policy on oil exploration must include the side effects on natural gas supply. For example, when the free trade in oil strategy could be proposed seriously, opponents argued that the benefits in lower oil import prices would be offset by the costs of higher natural gas import prices. Conversely, if protection stimulates additional oil exploration (the elasticity of this response curve is another question), the discovery of new gas reserves associated with this would have to be taken into account. On a more general level, the possibility of interfuel substitution means that higher oil prices raise the threshold level at which it becomes economical to exploit the marginal reserves of gas, oil shale, and coal.

Fifth, evaluation of the effects of changes in oil import policy is more difficult because the responses that follow are not pure market responses but the planned responses of a closely regulated and highly organized industry. An attractive proposal is to give the oil industry the degree of protection against imports justified in the national interest but to abolish all the other discriminatory measures in its favor, including expensing and treatment of royalties as income taxes. The abolition of percentage depletion for all but the small independent firms and the tightening-up of foreign tax credits are first steps in this direction.

5 Coal, Gas, and Electricity

Coal

The industrialization of the developed world, especially in the nineteenth century, was based upon coal. In the United States the coal industry had its origins in the eighteenth century with the mining and bituminous coal in Virginia and anthracite in Pennsylvania. By 1900 coal supplied 90 percent of US energy consumption. During this century, however, the contribution of coal progressively declined. By 1950 its share in energy consumption had fallen to 38 percent. The obvious explanation was the increased availability of more convenient and moderately priced fuels, namely domestic oil and natural gas, and the rapid development of energy users that could not use coal, e.g. automobiles. Although since 1950 coal output has remained fairly steady, its share in energy consumption has continued to fall dramatically (17 percent by 1972). This trend has been accentuated by several government actions: promotion of nuclear power, a direct competitor in the electricity utilities industry; the elimination in 1966 of oil import quotas for residual oil, which encouraged many large coal users to switch to foreign oil; and the uncertain prospects for coal as a fuel associated with the implementation of the Clean Air Act. Coal output has increased modestly since 1965, due to a rapid expansion in surface mining which is cheaper and less subject to escalating costs.[a] However, electric utilities have been the only expanding purchasing sector (purchases up 58 percent, 1965-73), while sales to industry and to retail outlets have slumped (by 35 and 50 percent respectively).

Recent market conditions in the coal industry have been at variance with the historical trend of decline. In 1973, for example, prices for large purchasing contracts—the most important type of sale—rose by at least 15 percent, and spot market prices soared. Despite the expensive wages contract awarded in 1974, there is no reason to expect costs of coal production to rise faster than the costs of obtaining other fuels. The sharp increases of 1973 and 1974 were due to special circumstances. Demand rose unexpectedly because of the Arab embargo, higher oil prices, and natural gas cutbacks. Supply could not respond immediately, and indeed there were few signs of bringing new mines into production.

---

[a]In 1973, coal was about 600 million tons, evenly divided between underground and surface mines. However, between 1965 and 1973 the output of surface mines had increased by 58 percent, while that of underground mines had fallen by 11 percent. The main reason was that underground costs had increased at double the rate of surface costs.

The lead-time for a mine is 3-5 years even if producers react quickly to a change in market conditions. In fact, very few new mines were opened in the early 1970s, and investment in related industries such as mining equipment producers and the railroads was low. The explanation is related to the fact that the capital outlay for a new mine, perhaps more than $25 million for a 3-million-tons-per-annum mine, can be recovered only over a period of 20-25 years. In spite of recently improved market prospects, the outlook remains uncertain because of the possible effects of strip-mining legislation, the implementation of the Clean Air Act, changes in oil import policy, pricing policies for natural gas, leasing policy in relation to western coal lands, and the unpredictability of nuclear capacity and electricity demand forecasts. These doubts have delayed investment.

In other countries coal mines are often under public ownership, so that many of these uncertainties are resolved via internalization of the external effects of energy policy decisions. This solution is hardly practicable in the United States in view of society's ideological resistance to such a move. However, the competitiveness of the US coal industry has been reduced in recent years, partly through the departure of many small coal operators but primarily because of a series of mergers and takeovers. In particular, the oil companies have acquired a substantial stake in the industry, accounting for one-quarter of production compared with only 2 percent in 1962. In addition, many large oil companies, such as Exxon, have obtained control of massive coal reserves. The problem with this concentration is that experience suggests that the opening up of new coal mines and the expansion of coal production by the oil holding companies will respond more to what is good for the oil companies' profits rather than to the energy supply needs of the American economy. In many situations the former goal may imply output restriction rather than the reverse.

The federal government ought to be able to play a more positive role in promoting supply, since it owns land containing 48 percent of the nation's coal reserves. Yet these lands produce only 2 percent of coal production, little more than ten million tons. Also, 85 percent of strippable low sulfur deposits are on public land. Hitherto, federal leasing policy has been too lax. There has been little government planning, coordination, or direction. Lands have been leased in response to private demand. Under the Mineral Leasing Act 15 billion tons of reserves are presently under lease, with a further 7 billion tons subject to prospecting permits. These lands have been leased at a negligible rent, typically $1 per acre, which hardly induces lessees to rush into production. There has been virtually no competition for leases: of 530 leases issued up to 1973, only 265 were "competitive" and of these 195 had only one bidder, often the owner of the adjacent land. The government has been handicapped in this respect by ignorance of the resource position, since land can be leased competitively only if "workable" seams exist in commercial quantities. Moreover, although terms can be reviewed after twenty years, in effect the leases are issued in perpetuity. The poor results in the form of coal output induced the Bureau of Land Management in 1971 to slow down the coal leasing program almost to a halt. Since reserves already under lease are equivalent of many years of current or projected coal output, the efforts of the federal government should be directed toward strategies to induce mining of this coal in ways that protect the environment, have favorable regional impacts, and meet energy needs, rather than by extending the coal leasing program.

The problem of accelerating coal supply has a critical regional dimension. As Table 5-1 shows, a substantial proportion of the national coal reserves—especially the low sulfur coal needed to meet environmental standards—is located in the western states, mainly in sparsely populated regions distant from the main consuming markets. Hitherto, these areas have contributed little to coal production. Given that coal mining, particularly surface mining, involves substantial environmental costs and that western residents are probably more environmentally conscious than those of the East, the development of western coal for national energy needs raises issues far beyond the narrowly economic. How serious this issue will become may depend on the degree to which coal is substituted for other forms of power and on the rate of growth in energy consumption.

The PIB base prediction (business as usual) for 1985 does not require a substantial switch to western coal (see Table 5-1). Substantial reserves still exist in the traditional Appalachian region and in the midwestern coal fields. The rate at which these can be exploited, however, is uncertain, because of the dominance of high sulfur coal. Much depends on where the line is held on air quality standards and on the satisfactory resolution of the stack gas scrubber (flue gas desulfurization) problem, including widespread adoption of the new technology when available and fully tested. An optimal production schedule aimed at meeting a predetermined high level of demand for coal at minimum costs (including transportation costs to market) would suggest initial concentration on low sulfur reserves in the East (particularly southern Appalachia), a second phase of developing western coal, and finally reliance on the high-cost high sulfur coals of the east, where costs are raised by the need to clean up the fuel mined. The need to include this third source is the consequence of the "clean fuels deficit problem" resulting from short-run supply inelasticities in low sulfur coal production. In the longer run, this difficulty can be resolved in any of several ways: the decision to develop western coals at a faster rate; elimination of the environmental bottlenecks to faster exploitation of eastern high sulfur reserves; or a reduced reliance on coal, presumably because of faster growth in nuclear power. It is not easy to assess at the present time which of these options is the most probable or the most satisfactory.

Any acceleration in the development of coal raises severe bottlenecks and obstacles, though none of these arises from a shortage of coal per se. Estimated coal reserves work out at over 820 years of current consumption, and even

Table 5-1
Regional Coal Statistics

Region	Reserves (b.tons)	Reserves Q. Btu	Mean Sulfur Content (% by wt.)	Output (m. tons) 1965	Output (m. tons) 1973	Output (m. tons) 1985 (BAU)
N. Appalachia	73.2	1922	2.0	191	177	243
S. Appalachia	39.1	1052	1.0	195	205	374
Midwestern	104.6	2492	3.1	121	150	227
Gulf	4.3	71	1.0	—	7	43
Northern Great Plains	175.4	3364	0.5	6	32	155
Rocky Mountain	23.7	568	0.5	13	24	39
Pacific Coast	13.6	262	0.2	1	4	19
National	433.9	9731		527	599	1100

Source: Federal Energy Administration, *Project Independence Blueprint* (Washington, D.C.: U.S. Government Printing Office, 1974), pp. 101, 103, 104, 108.

allowing for coal that cannot be mined with current technology reduces this estimate only to about 600 years. The constraints are of a different kind: safety, uncertainty about future demand, equipment, water, transportation facilities, land-use disruption.

Appalachia is still the dominant coal-producing region of the United States, accounting for three-quarters of output. Several problems are associated with coal mining in this region: reduction of groundwater resources; degradation of surface water quality due to acid mine drainage; soil erosion of strip-mined land; highwalls in hilly areas; land subsidence (control costs estimated at over $1 billion); and damage to miners' health. Two questions require a brief individual consideration: the profitability of strip mining if environmental controls are stringent; and how to improve the health and safety record without too adverse effects on productivity.

The reclamation problems associated with strip mining include spoil bank erosion and instability; water quality changes from acid mine drainage and similar geochemical reactions; revegetation; sedimentation; displacement of alternative or subsequent land uses; and secondary effects due to flooding or damage to roads, housing, and other properties and activities. The traditional method of assessing reclamation costs was to measure the first three of these items. Estimates by the Surface Mines Association and the Bureau of Mines suggested a modest cost of $300-500 per acre disturbed for these items. However, the full costs (i.e. adding on sums to cover the latter three items) are much higher, perhaps up to $1500 per acre. Since this implies a damage cost averaging $4 per ton, and this sum is far higher than average profits in the Appalachian strip-mining industry, the inference appears clear. If deep coal is a perfect substitute for surface coal, it would appear more efficient to abolish strip mining than to reclaim the land, including all associated social costs. In fact, this is an oversimplification. Variations in profit and costs around the average are very wide. Profits are highly sensitive to geological variables such as the density of the coal seam and the extent of the cover. Moreover, damage prevention costs (including full restoration) are much lower than reclamation costs; prevention costs depend *inter alia* on the method of strip mining employed.[1] Moreover, this analysis of the potential profitability of many but not all surface mining sites in Appalachia was based on 1973 coal prices. Since then, coal prices have risen, and will rise, much faster than damage prevention costs. Nevertheless, these arguments do not contradict the fact that the hilly terrain of Appalachia makes strip mining a much more costly venture than in other, flatter areas.

The health and safety record of underground mining in the United States is poor. The accident rate implies about eighty injuries, one-third of which are disabling and one is fatal, for every 4 million tons of coal mined. In 1973, several years after the passing of the Coal Mine Health and Safety Act of 1969, there were 131 fatal and several thousand other injuries reported in the US bituminous coal mines, which employ some 150,000 miners. The illness rate, measured as days of disability per million employee hours, is more than double the rate in other "unsafe" industries such as construction, metal mining and milling, lumber and primary metals. The chronic illness problem is possibly even more serious. The cases of "black lung" resulting from long-term inhalation of coal dust total about 125,000 and the disease is at least a contributory factor in 3000-4000 deaths per year. The 1969 Act set dust exposure limits, made provision for job transfers for sufferers, and introduced a black-lung benefits program (estimated cost $8 billion). If implemented rigorously, the act should eventually make a big difference. One consequence of the act has been a drastic fall in labor productivity, by 30 percent between 1969 and 1974.

Although other factors have been influential, such as an influx of untrained labor and poor labor-management relations, some argue that the economic cost of high health and safety standards is too heavy. However, it is possible to run safe mines efficiently. Accidents in European mines, such as in the United Kingdom, are less than half the rate experienced in the United States. More significantly, within the United States industry the intercompany variations are very wide. For example, in 1972 and 1973 the injury rate in Eastern Associated mines was 15-20 times greater than in United States Steel mines. The best mines have safety records that compare well with those in the most harmless service industries.

Outside Appalachia, the relative weight of problems changes. In the midwestern coal fields, for example, substantial surface mine reserves, flat terrain, and adequate rainfall make for easy reclamation (though this has not prevented the growth of 40,000 acres of unreclaimed land in Illinois). Nevertheless, a strong reclamation law requiring replacement of the topsoil would not create many problems in this region. The major constraint with midwestern coal is its very high sulfur content, for which the only solutions are technological. Another area where coal has a relatively high sulfur content is the Southwest, particularly the North West of New Mexico. Here gas companies serving southern California intend to construct coal gasification plants. Apart from high capital costs (over $4 billion), this project creates serious environmental problems, and would disturb the life styles of local, particularly Indian, populations. The offsetting benefit is an influx of new jobs (perhaps 14,000 miners and 5000 gas plant workers).

The most critical problems, however, relate to the development of western coal. Apart from the political question of the disruption of the western natural environment for the benefit of eastern industry and consumers, the major issues are the ecological effects of strip mining, the pressure on scarce water resources, and the impacts on the interregional transportation system. Although the coal reserves (lignite and subbituminous coal) of the Northern Great Plains region (Montana, Wyoming and the Dakotas) have a low Btu value (about 50 percent of Appalachian coal), the seams are thick with a shallow overburden, making strip mining attractive. But the environmental drawbacks are substantial: possible danger to wildlife; groundwater contamination because of limited groundwater recharge to eliminate polluted aquifers; destruction of open range and agricultural land, and other land-use impacts; uranium traces in some of the coal seams (perhaps 0.005-0.02 percent by weight); irremediable soil erosion; strains on the infrastructure and services of nearby communities. The light rainfall in many western states means that the soil cannot retain moisture so that revegetation after mining may be unsuccessful. This suggests careful selectivity of areas with moderate rainfall and good quality soil, such as the mixed grass region of the Northern Great Plains and the Ponderosa pine and mountain shrub zones of the Rockies. Strip mining can affect water supply for other uses, since the coal seams of near-surface coalbeds frequently trap underground water. Removal of these seams can disrupt these aquifers and diminish available water supply.

This influence on supply is paralleled by a more serious impact on demand. The development of western coal adds substantially to water demands, not so much in mining itself but in coal-burning power plants, coal gasification and liquefaction plants, and to cater for new mining and ancillary communities. Other than nondevelopment, two solutions have been proposed. One is to import water by huge long-distance aqueducts (as suggested in the USDI Montana-Wyoming Aqueduct Study). This remedy has serious objections. Making too much water available in these regions could induce rapid population growth in areas committed to low-density land uses such as ranching, outdoor recreation, and wildlife preservation. Aqueducts would be a severe disturbance to scenic amenity. Also, the solution is inflexible since interregional water transportation systems of this kind incur opportunity costs in the form of alternative uses for the water and the land. The preferable remedy is to separate mining from coal conversion by exporting coal to power plants or coal synthetics plants located in regions with adequate water supplies. The social and demographic impacts may then be minimized since mining and reclamation alone would involve much less development than if associated with coal conversion processes (with the latest technology a 5-million-ton mine can be operated with 120 men). This solution, of course, raises a further question—the adequacy of interregional transport systems and infrastructure.

The railroads are the dominant transportation mode for coal, with 78 percent of the traffic. Waterways are also important in some areas (15 percent of coal freight), but trucks are of little significance and merely cater for short hauls (the average haul is sixty-seven miles). A new mode is the slurry pipeline, but its use is limited by water availability. Where feasible, a slurry is competitive with the barge at volumes greater than 10 million tons per annum. Transportation for coal is a much more critical problem than for other fuels because of the expense, averaging 30 percent of delivered price compared with less than 5 percent for oil. The demands on the transportation system up to 1985 are not expected to be excessively heavy, because most projections anticipate less than full development of western coal before that date. If this assumption is correct, demands for new track will be very small. Upgrading of roadbeds, replacements of locomotives, and orders of new large hoppers should be sufficient. Steel requirements might be about 10.5 million tons, while capital requirements could be of the order of $11 billion. Although the industry has a poor profitability record (average rate of return equals 2.6 percent, 1961-73), it has been able to raise capital in the past, partly from depreciation allowances, partly via equipment debt. The waterway system could play an important role in delivering low sulfur western coal, combining routes with the railroad. The additional freight might be as high as 60-70 million tons per annum above current levels, requiring an investment of more than $700 million and more than 1 million tons of steel.

Other possible bottlenecks associated with the exploitation of western coal are the lack of availability of mining machinery and labor shortages in underpopulated areas. Neither is expected to be serious over the next decade, though temporary shortages could boost costs.

There is a high degree of uncertainty about the future demand for coal, both in the short and the long runs. The major shorter-term influences are the rate of conversion of oil and gas power plants to coal, and the division by fuel of new power plants—particularly the split between coal and nuclear power, and the degree to which coal use is restricted by environmental regulations. PIB estimated that switching existing power plants and large-scale industrial users to coal, prohibiting new oil- or gas-fired power plants, and encouraging home and commercial heating by electricity could substitute 400 million tons of coal per annum for 2.5 MBD of oil and 2.5 TCF of natural gas.[2] This maximum is unlikely to be reached. The MIT projection of coal demand for 1980 based on extrapolating recent consumption levels was 700 million tons, which might rise by another 75 million tons if power-plant conversion was maximized, requiring a switch of 44 percent of oil plants and 12 percent of multifuel plants.[3] The environmental constraints could be serious even if federal air pollution standards are relaxed, because many states impose higher standards than those prevailing nationally. This could preclude the use of about 225 million tons of coal now being consumed by the power companies. Without a low-cost desulfurization solution, demand could be met only by relying on the western coal, which though cheap to mine is in inelastic supply in the short run, and may become increasingly expensive because of reclamation and transport costs. From the 1980s on, the main determinant of coal demand may be the degree of success in synthetics technology, though few expect synthetics to provide more than 10 percent of energy needs even by the year 2000.

How much coal will be used in the longer-term future is very much a wild guess, and hence the eastern-western supply ratio is also hard to forecast. PIB's forecast for 1980 was higher than that of MIT (895 million tons as opposed to 800 million tons) even under BAU conditions, and could reach 1376 million tons if accelerated supply strategies were pursued. Its range for 1985 was 1100-2063 million tons, with a western share little higher than 20 percent.[4] The

Ford Foundation study argued that the range could be even wider, 640-1120 million tons for 1985 and 680-2120 million tons for the year 2000, depending upon the set of assumptions chosen.[5] The lower end of these forecasts implies little strain on western coal supplies. For instance, the Ford Foundation's environmental protection version of its technical fix scenario, even assuming a western 50 percent share of the 1973-85 increment in coal production, implies only a 25 percent increase in western coal output over this period and a surface mining rate of only ten square miles per annum. Even taking a high projection for the year 2000 (say 2 billion tons per annum), the 50 percent share for western fields would require no major extension in the coal leasing program.

Capital investment in coal mining varied between $300 and $700 million per annum over the years 1966-73. Expansion in coal output over the next decade will raise investment requirements so that, on the assumption of a capital cost of $15 per ton in surface mining and $20 per ton for a deep mine, $12 billion could be needed by 1985. (The range of estimates is $6 to $18 billion, at 1973 prices.) This should not present serious problems, in view of the oil industry's heavy involvement in the coal industry. The financial resources of the electric-utility industry could be a more severe constraint. On the other hand, there might be a danger of building too many coal-fired power plants. Demand for coal will primarily manifest itself over the next decade in increased electricity consumption, whereas from about the mid-1980s an increasing share of coal demand will probably be related to liquefaction and gasification.

On the whole, therefore, detailed examination of the evidence supports the generalization that increased reliance on coal is demand rather than supply constrained. The major problem is the interdependence of decisions and their timing. Wise decisions are more likely to be taken within a context of a *comprehensive* and fully coordinated energy policy rather than ad hoc responses to market price triggers. At the time of writing, there is little evidence that such a policy is being developed.

Natural Gas

Natural gas has many advantages over alternative fuels. It is usually sulfur free or easily purified, and it can be burned completely in many different types of appliances and apparatus. It leaves no residue, and is a very low pollutant. It is hardly surprising that it has accounted for an increasing share of energy production and consumption: its share in production rose from 11.9 to 40.6 percent, 1940-73, while that in consumption rose from 12.4 to 34.6 percent over the same period. It is not expected to maintain this share, however. Recent forecasts suggest that its share in energy inputs might drop to below 18 percent by the year 2000. Unless some dramatic discovery is made, proven reserves are very low (about 270 trillion cubic feet), not much more than a decade of consumption at current levels.

Estimates of probable undiscovered reserves suggest a total of about 1200 trillion cubic feet, implying a reserve-production (R/P) ratio of about 50. This is less than appears at first sight. First, consumption has been increasing rapidly and will continue to do so, unless controlled by price increases, Federal Power Commission (FPC) regulations, or other devices. Second, of the undiscovered reserves most are more difficult to extract than in the past: 14 percent are at depths greater than 15,000 feet, more than 20 percent are offshore, and 28 percent are in Alaska. This last point is quite important, since one disadvantage of natural gas is that it is much more expensive to transport than oil. Natural gas pipelines have to be four times the size of oil pipelines to carry an equivalent amount of energy, while liquefied natural gas tankers have to be double the size of an equivalent oil tanker. Thus, the remoteness of natural gas fields from markets can be a critical consideration in their attractiveness.[b]

If the long-run prospects for natural gas seem less than bright, the short-term situation, at least in the recent past, has been critical. Shortages have been acute. Natural gas contracts to customers with firm contracts have been curtailed in successive years since 1971. In winter months since 1970 gas supplies have frequently been cut off to industry and to residential and commercial consumers in many northern and eastern states. The shortages have been more acute, if less visible, in long-term contract and reserve markets than in the current gaps between production and potential consumption. Estimates of the shortage varied, but all were substantial. Even the FPC predicted a shortfall of over 12 percent in 1975 and an actual shortfall of over 5 percent in 1972. This does not take account of buyers who were refused service and those who had to turn to alternative fuels. MacAvoy and Pindyck[6] estimated that potential demand at the prevailing price exceeded the total supply available by more than 50 percent. Erickson and Spann[7] estimated that for an earlier date (1968) the "real shortage" (i.e. the gap between actual supply at the prevailing price and demand at the *market* price) was between 10 and 25 percent. Since they also argued that the price elasticity of supply of new discoveries is about +0.5, this implied that a price increase of 20 to 50 percent would be necessary to clear the market.

There is ample other evidence of the shortage. Additions to reserves have fallen below annual production in every year since 1968 apart from 1970 (Table 5-2), so that the R/P ratio has been declining dramatically. Even if additions to reserves keep pace with production, the R/P ratio will decline if the trend of

---

[b]An interesting study (L. Waverman, *Natural Gas and National Policy: A Linear Programming Model of North American Natural Gas Flows*, Toronto: Toronto University Press, 1973) shows that Canadian consumers incurred much higher costs because natural gas was brought from western to eastern Canada by pipeline rather than being exported to the western states of the United States in return for imports into Ontario and Quebec from Texas, Louisiana, and Kansas, which are nearer. In other words, the natural (i.e. free trade) routes were on north-south, not east-west lines. Needless to say, the climate for a transition to free trade between the United States and Canada in energy resources is much more unfavorable today than in 1966, the year of the data used in Waverman's study.

Table 5-2
Natural Gas Reserves, Production and Price

Year	Reserves (tr. ft.^{3})	Discoveries (plus Extensions) (tr. ft.^{3})	Production (tr. ft.^{3})	R/P Ratio	Av. Field Price (Cts/Mcf)
1960	262.3	13.9	13.0	20.2	15.53
1961	266.3	17.2	13.4	20.0	16.30
1962	272.3	19.5	13.6	20.0	16.54
1963	276.2	18.2	14.5	19.2	16.63
1964	281.3	20.3	15.3	18.4	16.55
1965	286.5	21.3	16.3	17.6	16.66
1966	289.3	20.2	17.5	16.5	16.70
1967	292.9	21.8	18.4	15.9	16.96
1968	287.3	13.7	19.4	14.9	17.20
1969	275.2	8.5	20.7	13.3	17.48
1970	290.7	37.5	22.0	13.2	17.93
1971	278.8	10.6	22.5	12.4	19.13
1972	266.1	9.8	22.5	11.8	20.54

Source: American Gas Association, *Gas Facts, 1973* (New York: American Gas Association, 1973).

production is upwards. In this case the *absolute increase* in additions to reserves from year to year must be many times greater than the absolute increase in production. The drilling rate—the dominant factor determining future supplies—has declined perilously over the last fifteen to twenty years. The rate of decline in drilling footage was about 4 or 5 percent per annum[8]; the number of wells drilled fell from about 57,000 to 30,000 a year over the period 1950-70. Imports of natural gas, predominantly from Canada and Mexico, have increased rapidly from 11,000 Mcf in 1955 to over 1 trillion cf in 1973, over a period when exports have remained insignificant at less than 8 percent of imports (see Table 1-4, p. 14). Negotiations are under way to boost imports of natural gas (in the form of liquefied natural gas) from more distant sources—Venezuela, Algeria, and even the USSR.

Why has there been such a severe shortage? There is a simple explanation based upon elementary price theory. This is quite valid, but it requires amplification in terms of the characteristics of the natural gas industry and the nature of its regulation by the FPC. The shortage—an excess demand for natural gas at the prevailing price—was due to the price being too low to clear the market. The price was too low, but it was compulsorily regulated at that level by the FPC. The simple solution to the shortage would be to deregulate the price and allow it to rise until it clears the market. The objection to this clear-cut and obvious solution is that supply is so inelastic in the short run that market clearing is achieved only at the expense of a massive price increase. This may be true. The counterargument is that the benefits of price adjustment cannot be assessed in the short run, primarily because the merits and demerits of price regulation depend upon the impact of reserves. A higher price, it is argued, will stimulate more exploration and discovery so that after a lag, perhaps five years, possibly ten, supply will be increased markedly. Demand will also increase over time, but it is suggested that the rightward shift in the supply curve will be greater than that of the demand curve. If this were so, the market price in 1980 would be lower than in 1975. Although it may be higher than the regulation price that could prevail in that year if regulation was continued, consumers would benefit because they would obtain the supplies needed to satisfy demand. Unsatisfied demand is a cost to consumers that may be more burdensome than higher prices.[9]

This argument appears very strong, but it would be wrong to accept the face value of an abstract, simple model without considering the real world characteristics of the industry and how it has been regulated.

Several features of the industry are important in this context. As an exhaustible resource, natural gas is a commodity that is sold in future as well as present markets that are closely interlinked. Expectations about the future price level due to estimates of future resource availability and future demand will directly influence the present price level in an unregulated market. The R/P ratio may be a crude rule-of-thumb guide to these expectations. If the R/P ratio falls below the desired (or equilibrium) level, it will be taken as an indicator of future price increases and supply will be sustained in the present only if the price is allowed to rise. The appropriate price adjustments will occur in an unregulated market because of an idiosyncrasy of the natural gas industry—namely, that gas producers sell to pipeline firms in the form of long-term contracts, usually twenty years. Accordingly, new contract prices will reflect producers' expectations of future price changes. In the regulated market, however, the price is simply a function of cost, set on a cost-plus-margin basis, and ignoring the long-term reserve position. In the free market, low R/Ps will be reflected directly in higher contract prices.

Another effect follows from the contract feature of the natural gas market. The impact of gas price changes on the economy is felt only sluggishly, because of the "rolled-in" effect. An increase in the wellhead price affects wholesale prices only to the extent that new contact prices change the historical average of all field prices. (Average contract prices have tended to be a few cents below the new contract price. The price increases of 1974 widened the margin considerably.) The full impact occurs only after several years, depending upon the proportion of contracts being renewed each year. This makes the brunt of a deregulation move much easier to bear.

A price increase may induce two different kinds of exploration response.

Intensive drilling involves drilling where gas has already been discovered; *extensive drilling* means the riskier exploration of new fields, usually offshore or onshore at much greater depth. There is a trade-off between the probability of discovery and the average discovery size on initial drilling, and it is not possible a priori to forecast how price changes will affect the intensive-extensive distribution. However, large discoveries are worth proportionately more than small because of substantial long-term economies of scale in long-distance pipelines. The greater the price increase, the more likely the shift to extensive drilling, with its greater risks but also greater payoffs because of scale economies. Also, on reasonable assumptions about the existence of gas and about success ratios, the greater the elasticity of supply. Khazzoom[10] showed not only that long-run elasticities were high, but also that they increase with the size of the price increase, though they converge towards a limit that depends on initial conditions.[c] The converse of this argument, of course, is that an artificially low regulated price can have a disastrous effect on the rate of new discoveries. Deregulation, on the other hand, and its associated, possibly substantial, upward shift in price, may have a dramatic effect on new discoveries—though these will become effective only after a time lag because of lead-times. Yet the impact of the higher price level on the economy is mitigated because of the "rolled-in" effect.

The argument is persuasive. However, it should be recognized that there are other ways to stimulate new discoveries, apart from price increases. One suggestion is to replace the leasing system for lands suspected of containing gas or oil by a delayed bonus payments system, under which the producer pays a proportion of the lease sum, say one-fifth, when the lease is acquired, another fifth upon discovery, and the remaining three-fifths when production gets under way. If wells are dry, the producer is exempt from four-fifths of the payment. This type of strategy reduces the risks of discovery considerably. Other schemes of this kind are not difficult to devise.

The price structure of the natural gas industry is complicated because there are several interrelated markets under varying degrees of control. There is a field market for reserves in which gas producers commit new reserves to pipeline companies under long-term contracts. Then there is a wholesale market for gas production in which the pipeline companies sell gas to retail utility companies and to industrial consumers. These are interdependent because demand in the wholesale market helps to determine the pipeline companies' demands for gas reserves in the field market. The FPC began to regulate the prices charged by the pipeline companies in 1938 long before it regulated the field price (after the

---

[c]The most important of these initial conditions are the level of initial discoveries and the base price. Although positive, the price elasticity of gas may not be as high as industry spokesmen sometimes suggest. The price elasticity of total discoveries is the sum of the price elasticities of wildcat drilling, the success ratio, and average discovery size. The former is positive, but the latter two are negative.

1954 Supreme Court decision in *Phillips Petroleum Co. v. Wisconsin*). The latter, however, has been much more effective since the early 1960s; the field prices were held down throughout that decade well below the rate of increase in the general price level (Table 5-2, p. 81). A tight rein on the field price has depressed supply, while a much less effective degree of control in the wholesale market has prevented consumers from obtaining maximum benefits from lower prices (perhaps this was fortunate; otherwise the excess demand might have been much worse). However, FPC control applied only to interstate shipments, and intrastate sales have remained uncontrolled. There has also been a distinction between new contract prices and old prices; the FPC's price hikes apply to new contracts and affect the old only as they are renewed.

Moreover, the FPC's pricing principles have changed over time. In 1960 (as a result of the Permian Basin decision) the concept of area-wide ceiling prices replaced the principle of pricing by cost of service. These ceiling prices were based on the concept of average cost. They referred to pipeline quality gas. In effect, the FPC calculated a national average cost of finding, developing, and producing gas. To this national average were added regional costs, quality improvements, royalty charges, plant processing costs, gathering costs, state production taxes, and an allowance for a fair rate of return on investment. The price does not vary with volume, so that it represented, in effect, the average and marginal cost of gas to the interstate pipeline company. However, because of tardiness in agreeing rates, the area price ceilings in most areas did not exist, and substitute rates, called in-line and guideline rates, had to be used. The process of determining area rates for the country as a whole had not proceeded very far when in June 1974 the FPC replaced the incomplete area price structure with a much higher national price (43 cents per thousand cubic feet). (The national price applies only to the gas fields; the price in some non-producing states was three times the regulated price.) This was partly in response to the criticisms about its slowness of operation, partly a response to the dramatic shift in oil prices after 1973. However, despite a further increase in December 1974 the relationship between gas and oil prices was still heavily in favor of oil. Accordingly, the excess demand for gas still remained. The absolute price increases that had been allowed provided some incentive to exploration, but resources are much more likely to be devoted to searching for oil.[d] The justification for con-

---

[d]To some extent, natural gas and oil are joint products in exploration, development, and production. If gas prices are too low relatively to oil prices, exploratory effort is concentrated on discovering oil. In this case, a high proportion of new gas supplies is "associated gas." On the other hand, some analysts attach little importance to this directionality factor on the ground that wells cannot be identified *ex ante* as either gas or oil wells.

Higher oil prices have offsetting impacts on gas supplies. In the short run, there will be an increase in the search for oil at the expense of the search for gas. In the longer run, increased investment in oil discoveries will have a favorable effect on gas supplies because of discovery of gas reservoirs as a by-product of the intensified search for oil. It is impossible to

tinued price regulation becomes very weak. The costs of regulation are high, and the benefits dubious for producers and consumers alike. It would be preferable to allow the market to determine the price of natural gas, and for the FPC to adopt a "watchdog" function to ensure that competitive rather than monopolistic forces prevail. This strategy would avoid the now widely recognized mistakes of recent years and result in an improved distribution of supplies and demands among substitutable fuels. At the time of writing, deregulation is under consideration as part of a policy package to stimulate domestic energy supplies.

Electricity

The growth of the electricity industry since 1950 is illustrated by the data shown in Table 5-3. Generated energy and sales to consumers have increased approximately sixfold since that date. Residential and small nonresidential consumers have increased their share of demand relative to the large-scale commercial and industrial customers, partly via an expansion in the number of consumers but primarily because of much higher electricity consumption per capita (by more than fourfold, 1950-73). Part of this growth was stimulated by low-cost energy, since fuel bills over most of this period did not rise as fast as consumption. However, between 1970 and 1973 average bills per customer increased by over 30 percent,[11] and since then unit electricity charges have been escalating even faster.

As Table 5-3 shows, generating capacity has increased more than sixfold over the period since 1950. Prior to 1970 much of the increase was due to fossil-fueled steam plants, since the growth in hydroelectric power capacity has been steady but unspectacular. Since 1970 nuclear power stations have become a significant element in electricity supply. Although its share remains small—about 5 percent of total capacity—this could change drastically in the near future.

The pluralistic ownership structure of electricity supply in the United States is unique. There are about 3500 electric power systems, of which almost 30 percent are engaged in generation and transmission. Most of the power—more than 90 percent—is generated by 150 or so companies. The federal government is also an important supplier, particularly via the TVA (Tennessee Valley Authority) and several Department of the Interior agencies. Municipal systems account for most of the 2000-plus public power operations. In addition, the Rural Electrification Administration (REA) controlled about seventy generating co-

---

determine a priori which of these effects is dominant. If the directionality factor is weak, however, the net effect should be favorable.

If gas prices are held below their equilibrium level while the price of oil clears the market, the low gas prices will tend to shift the supply curve for *oil* to the left, and this will reduce the discoveries of associated gas. The joint cost aspects of the search for oil and gas aggravate the initial gas shortage.

Table 5-3
Electric Power Industry

	1950	1960	1965	1970	1973
Energy generated (b. Kwh)	329	755	1,055	1,532	1,849
Sales to ultimate consumers (b. Kwh)	281	683	953	1,391	1,703
Residential and domestic (%)	23.8	28.7	29.5	32.2	32.5
Small commercial and industrial (%)	17.8	16.8	21.2	22.5	23.3
Large commercial and industrial (%)	49.5	50.5	45.4	41.2	40.3
No. of ultimate customers (m.)	45.0	58.9	65.6	72.5	78.5
Av. Kwh per customer ('000)	6.9	11.7	14.7	19.4	22.0
Residential	2.0	3.9	4.9	7.1	8.1
Small commercial	10.4	17.0	27.4	40.0	47.6
Av. annual bill ($)	123	198	234	308	408
Residential	57	95	111	148	192
Small commercial	265	418	584	803	1,094
Generating Capacity (m. kw)					
Steam					
Fossil	49	132	188	275	311
Nuclear	0	3	1	6	21
Hydro	18	32	44	55	67
Total	69	168	236	342	432

Source: U.S. Bureau of Census, *Statistical Abstract of the United States, 1974* (Washington, D.C.: U.S. Government Printing Office, 1974), pp. 518, 524.

operatives and nine hundred cooperative distribution systems. Thus, the public sector accounts for a great many of the small-scale operations that characterize the United States electric-power industry.

The evidence in support of rationalization of the industry has been supplied by Olson (1970), among others.[12] For both coal-fired and noncoal plants per kilowatt hour costs fell with increasing size of generator unit and with higher utilization. Transmission costs decrease with higher voltages, and there is a load level which minimizes per kwh cost given distance and voltage. Scale economies in transmission are substantial because load-carrying capacity increases in proportion to the cross-sectional area of the wire, while line cost increases in proportion to the diameter of the wire. Furthermore, an interconnected system permits substitution between generating and transmission costs. An integrated transmission network reduces generating costs by permitting the electric utility to benefit from scale, utilization, and location advantages. The adoption of large-scale generating units with strong transmission links would result in greater efficiency and lower electricity costs to consumers. Moreover, the dynamics of the situation under conditions of changing technology reinforce economies of scale, and render obsolete the local monopolies and isolated markets of the early 1900s and the schemes promoted by the TVA and REA in more recent decades. Installations in recent years have shifted in favor of larger units, and the evidence suggests that plants of below 100 megawatts are suboptimal.

While the FPC has recognized the advantages of large-scale operations—for instance, arguing that 1000 megawatt units might be the most efficient by 1980—it has done little to rationalize the structure of the industry.

Despite the long period of nominal responsibility for electric power, the FPC did not attempt to regulate the industry seriously until the 1960s. The National Power Surveys of 1964 and 1970 showed that close coordination of plant construction and operation in the nation's electric companies could result in much cheaper power, while there are considerable advantages to be gained from improvement in the national transmission system. The ability to plan effectively for the industry as a whole is an important criterion for evaluation of regulation in view of the adverse or neutral effects of price controls. Although coordination has been achieved in some respects, for example, power companies in New England and in New York have organized central dispatching, there are twenty-two regional pools, and a National Electric Reliability Council (NERC) organized into nine regional councils, the rate of progress has been slower than might have been hoped for in view of the potential gains (estimated at about $2 billion, or about 3.5 percent of total 1980 operating costs, plus the substantial environmental benefits from doing without unnecessary capacity). The Commission used exhortation and requests for information, but failed to resort to its stronger powers (e.g., requiring firms to interconnect). When it did take firmer action—such as after the Northeast power blackout of November 1965—this was in response to emergencies rather than as part of a long-term plan. The obvious solution to the difficulties and the one adopted in many other countries—public ownership—is probably too radical and too tinged with ideological overtones for American tastes.

A very recent problem is the reaction of the electric utilities to their recent liquidity squeeze and the current stagnation in electricity consumption (demand grew by only 0.6 percent in 1974). They have cut back their construction budgets, and almost 50 percent of planned new capacity has been cancelled or postponed (perhaps two-thirds of these cutbacks refer to nuclear power plants). The FEA is worried that these cutbacks could result in serious electric power shortages by the early 1980s. One problem is how to raise capital. Historically, about one-third has been raised internally, but in the last eight years the long-term debts of the utilities have doubled. Another is how to moderate peak demand growth, which has been, and still is, growing faster than demand as a whole. An optimistic solution is that by the 1980s technological changes may permit the utilities to build huge storage batteries. The rationalization of the industry discussed above could also alleviate the risk of insufficient capacity. If an effective Grid Act could be introduced, even if it co-ordinated transmission and capacity-sharing only on a regional basis, further capacity requirements might be reduced by 10 percent.

6 Nuclear Power

Introduction

In view of the possible scarcity of fossil fuels in the future, the question of an alternative source of power is important. Of the four main alternatives—solar energy, geothermal energy, and hydrogen and nuclear power—only the last is feasible at present as a major energy source on grounds of cost, flexibility, and known technology. In fact, nuclear power represents at least three very different forms of energy supply: inefficient burner reactors (the present form), which could conceivably use up the scarce resource U-235; fast breeder reactors, which create more fuel than they consume, and are the main hope of the next decades; and fusion reactors, which are scientifically possible but at present beyond technological capability.[a] If power from nuclear fusion became feasible and controllable, it would virtually free the world from the constraint of physical energy resources. In economic terms it would represent resource-saving technological progress and substitution of capital investment for resources. The significance of breeder reactors and fusion reactors to the problem of resource exhaustion is that breeders would represent a 130-fold increase in energy resources, while fusion would imply a 48,000-fold increase in resources over the level associated with breeders.[1] These estimates are only approximate, but they suggest that the world's energy problems are a mix of technological, economic, environmental, and social factors, not physical.

This is an economic rather than a technological analysis. Accordingly, only the briefest and simplest nontechnical explanation of the principles of fission and fusion technology will be presented. The major economic-related issues are of a different nature. They include:

1. the rate of use of U-235 prior to the widespread adoption of breeder reactors;

2. the cost of nuclear power relative to that of major fossil fuels;

3. the rate of substitution of nuclear for other fuels in electric-power generation;

4. the significance for resource economics of nuclear power as a *backstop technology*;

5. the evaluation of R and D investment in nuclear technology;

---

[a]However, an experimental fusion test reactor is to be built at Princeton University at a development cost of $215 million. The hope is to have this operational in the 1980s, with perhaps a 500-Mw demonstration power plant by the 1990s.

6. the health hazards of nuclear power relative to those of fossil fuels;
7. the cost of disposal of radioactive wastes;
8. other environmental and social issues related to nuclear energy (some of which are discussed in Chapter 10).

Nuclear Reactors

Uranium consists of three isotopes, U-234, U-235, and U-238. Their distribution is 0.006 percent, 0.711 percent, and 99.283 percent respectively. U-234 is too scarce and U-238, though plentiful, is nonfissionable. In its initial phases, fission reaction depends upon the isotope U-235, the only one of several hundred natural isotopes that is spontaneously fissionable by neutron bombardment.

The nuclear reactor is a physical construction in which controlled chain reactions occur. Reactors are of three main types: burners, converters, and breeders. Burners consume U-235 directly, and are wasteful because they consume this scarce resource at too fast a rate. Converters are so named because they are capable of converting nonfissionable but fertile materials, particularly U-238 and thorium-232, into isotopes that are fissionable, namely plutonium-239 and uranium-233, by exposure to neutron bombardment. The distinction between conversion and breeding is that whereas a converter only uses up a fraction of the fertile material before the supply of the fissile material is exhausted, a breeder produces more fissile material than is consumed, so that it is possible to use the entire supply of fertile material provided that enough U-235 is available to initiate the process. Since U-238 is 141 times more plentiful than U-235, breeders represent a very large expansion in the scope for nuclear energy. The most pressing problem in the nuclear-power field is to introduce breeder reactors soon enough and on sufficiently wide a scale to replace high-U-235-consuming burners (e.g. light-water reactors) and low-ratio converters before too large a proportion of low-cost U-235 resources is used up.

The major types of reactors, either in use or at the design stage, can be divided into two categories, nonbreeders and breeders. The main types of non-breeders are light-water reactors, either the pressurized water reactor (PWR) or the boiling water reactor (BWR), and high-temperature gas-cooled reactors. The breeder possibilities include, in addition to the well-known liquid-metal fast breeder reactor (LMFBR), the molten-salt breeder reactor (MSBR) and the gas-cooled fast breeder reactor (GCFBR).

One of the factors increasing the uncertainties about the rate of growth in nuclear power consumption is the wide range of technological choices, particularly in respect to the crucial breeder program. Although the competition among alternative reactor types may result ultimately in a better product, the short-term effects are dissipation of the research program and the impossibility of forecasting both the adoption date and the rate of diffusion.

The light-water reactors currently dominate the nuclear-power-plant field, reflecting to a large extent past technology and the fact that reactors have been installed since the late 1950s. These reactors use U-235 as fuel, and have—despite recent technological advances—relatively low conversion ratios.[b] The light-water reactors use ordinary water both to moderate the neutrons and as the medium by which heat is removed from the reactor. In the PWR the water does not boil but is pumped at a high temperature through a heat exchanger to make steam in a secondary circuit. This steam drives a conventional turbine, which turns the generator. In the BWR the water inside the reactor boils and its steam runs the turbine directly. Both reactors have a low thermal efficiency for converting heat into electricity (32 percent).

The high temperature gas reactor uses solid graphite as a neutron moderator and high temperature helium to boil the water for the turbine. It has certain advantages over the light water reactors: a higher thermal efficiency (about 40 percent) a better conversion ratio (more than 0.8), and possibly lower accident risks. On the other hand, there are teething problems in the fuel reprocessing techniques, and the makeup nuclear fuel is the expensive and perhaps vulnerable pure U-235.

Since about the middle 1960s the breeder reactor-development program has become the focal point of the Atomic Energy Commission (AEC) effort.[c] This was in response to a realization that energy scarcities might promote a faster rate of adoption of nuclear power stations than had been expected previously, and to the (challengeable) belief that breeder reactors would be safer, more reliable, and above all, more economic.[2] The favored type is the LMFBR using the U-238 → Pu-239 cycle. The nuclear fuels—both plutonium and uranium—are immersed in a large tank of liquid sodium metal. The liquid sodium is circulated through the reactor, and takes the heat to a succession of heat exchangers, where eventually steam is generated to power the turbines. Prototypes are being tested or developed in the USSR, UK, France, and the USA. The United States intends to build a 3000-Mw demonstration plant near Oak Ridge, Tennessee, to be completed by 1980. The AEC expects LMFBRs to be introduced commercially by the late 1980s and on an increasingly wide scale. The LMFBR has many advantages: a 40 percent-plus efficiency, low-pressure operation reducing failure possibilities; and it is a breeder. Nevertheless, its superiority over alternative reactors is still unclear.

Since the MSBR uses a thorium-232 → U-233 cycle, with thorium as a raw material, it could be a complementary development to the LMFBR. The GCFBR

---

[b]The ratio of fissionable fuel made to that used is about 0.6. Thus, the most up-to-date of these reactors regenerate some nuclear fuel to make plutonium via absorption of neutrons into U-238.

[c]The research and development activities of the AEC have now been absorbed in the Energy Research and Development Administration (ERDA), while the safety and licensing operations have been transferred to a new agency, the Nuclear Regulatory Commission (NRC).

is an extension of the high-temperature gas-cooled reactor; its promoters argue that its capital costs would be lower than those of the LMFBR. Another, probably less attractive alternative is to add blankets of fertile materials to a light-water reactor to increase its conversion ratio drastically.

Although the costs and the eventually favored reactor type are still unknown, it is probable that breeder reactors will be introduced commercially before 1990. The adoption rate depends upon the extent of breeding. For instance, a doubling time of ten years (i.e. the time required for the doubling of the initial fuel inventory), assuming no uses of fissile material for nonbreeding purposes, would allow a maximum rate of growth of breeder-power production of about 7 percent per annum. It is hoped that the LMFBR would achieve a ten-year doubling time, with a ten-to-twenty-year doubling time for the MSBR.

Fusion Reaction

The energy radiated from the sun and the stars is produced by the fusion of hydrogen of atomic-mass 1 into helium of atomic-mass 4 (as demonstrated by the subsequent Nobel prizewinner H.A. Bethe in 1939). Hydrogen has three isotopes of mass numbers 1, 2, and 3, known as hydrogen (H), deuterium (D), and tritium (T) respectively. Controlled fusion depends upon fusing two or more of these isotopes into helium (which also has two isotopes, helium-3 and helium-4). The most hopeful approaches are via the fusion of two deuterium atoms or of one deuterium and one tritium atom. The latter is perhaps the best bet, at least initially. Tritium is not found in nature, but is produced from lithium.

The fusion of deuterium and tritium into helium in an *uncontrolled* form has been achieved in the thermonuclear, or hydrogen, bomb. Although research into controlled reactions has been under way for over two decades, it is still not possible to estimate when, or even if, the development of power from fusion reaction will be achieved. Nuclear fusion reactions require heating the reactants to an ignition temperature of 100-1000 million degrees (much hotter than the center of the sun), and confining the "plasma" long enough for some of the nuclear fuel to "burn." How to extend the duration of the confinement long enough has absorbed most of the research funds. It is conceivable that the development stages will last until the end of the century, and the feasibility of the method is still in doubt. Moreover, scientific feasibility is only the first phase of a process that includes technological development followed by economic and social acceptability. Thus, fusion cannot yet be considered a viable alternative to the breeder program.

Even though it is impossible to evaluate the feasibility of controlled fusion reaction, it is possible to forecast its resource implications if and when developed. This is because the possible fusion reactions and their associated energy releases are known, and quite a lot is known about resource availability even though the search for lithium has not been pushed very hard hitherto. On the basis of known lithium reserves in the United States, Canada, and Africa the energy obtainable from fusion of lithium-6 is about 2.4×10^{23} joules, broadly equivalent to the 2.6×10^{23} joules energy equivalent of the world's initial supply of fossil fuels. Thus, it is only from the possibly more technically difficult deuterium-deuterium reaction that a complete escape from resource constraints in nuclear fusion can be obtained. Deuterium is in plentiful supply in water:

The fuel equivalents of 1 cubic kilometer of sea water are 269 billion tons of coal, or 1,360 billion barrels of crude oil. . . . The total volume of the oceans is about 1.5×10^9 cubic kilometers. Should enough deuterium be withdrawn to reduce the initial concentration by 1 percent, the energy released by fusion would amount to about 500,000 times that of the world's initial supply of fossil fuels.[3]

The Backstop Technology?

In the context of the controversy about the possibility of exhaustion of energy resources, nuclear breeder reactors are the most probable candidates for the role of the *backstop technology*.[4] This is defined as a set of processes having two key characteristics: it is capable of meeting demand requirements; and it has a virtually inexhaustible resource base. If the technical problems (including those of waste disposal) associated with breeder fission can be solved in the next decade or so, then the doomsday arguments about the exhaustion of energy supplies fall to the ground. The predictions of compulsory zero growth revolve around whether technological change can offset imminent resource exhaustion and the degree to which capital can be substituted for other resources. Breeder technology represents a good test of these critical questions. Current fuels are scarce and increasingly dear, but production costs are low; if the breeder problem is solved, nuclear fuels are relatively plentiful and cheap, but production costs are high.

If the technological problems can be solved in time, the question of the rising costs of energy resources ceases to be a matter of major importance. Sharply rising costs merely bring forward the switch date to the backstop technology. True, widening differentials in relative prices between energy resources and other goods and services may force economic and social readjustments on an economy, and induce far-reaching substitutions, but these merely affect the type of structural change occurring in the economy rather than determining the outcome in the zero growth debate. In this context breeder fission is not just another addition to the range of energy fuels, but rather changes the character of the problem. The responses to the prospect of imminent resource exhaustion are very different to those encountered in a situation characterized by having to choose the distribution of energy supplies from among competing fuels, the future relative costs of which are uncertain.

Nuclear Power Production

The growth of the nuclear power industry over the last decade or so is reflected in the data of Table 6-1. Although nuclear generating capacity remains a very small part of electricity generating capacity, the industry has already absorbed huge amounts of money—much of it out of public expenditure. Presumably the payoff lies in the future. Plant investment in excess of 10 billion dollars, annual AEC appropriations of 2.5 billion dollars, and employment of more than 112,000—the nuclear industry is clearly big business, even if there has been a slight relapse since its peak in the 1960s. Reactor development costs have been rising steeply, primarily because of the massive R and D commitment to the breeder reactor (especially the LMFBR). However, despite frequent criticisms of the AEC's policies, safety costs have also risen sharply, accounting for an increasing share of development costs (17 percent in 1973 as against 10 percent in 1965). Although the uranium ore reserve position improved markedly in the late 1960s, it has not altered much since 1970. However, this is probably explained by the fact that ore prices have not been high enough to provide a strong incentive to exploration (until recently) rather than by imminent resource depletion. Ore milling capacity has also not increased much since 1970, and this—in conjunction with enrichment plant constraints—could be more of a problem than ore reserves or even the price of uranium.

The nuclear-power industry is so embryonic that it makes little sense at this stage to attempt to forecast its long-term growth. At present, the United Kingdom produces about 10 percent of its electricity with fission and is the country most dependent on nuclear power. In the United States the comparable figure is 4 percent. "After more than $3 billion worth of development, nuclear power has only just surpassed firewood as an energy source."[5] Future production depends on so many variables: future energy demand, the fossil-nuclear split—a complex factor reflecting variations in relative costs and a myriad of other influences, the volume of R and D expenditure, the sequence and consistency of government decisions and other political pressures, the timing of major technological breakthroughs in, say, breeder reactors and radioactive waste disposal. Any forecast made now for the year 2000 is almost certain to be wrong. Nevertheless, it is of some interest to refer to some of the tentative projections that have been made.

First, it is clear that within this forecast period almost the sole use for nuclear energy will be in electric-power production, which at present absorbs only about one-quarter of total fuel consumption in the United States even though it has

Table 6-1
Nuclear Power Statistics

	1965	1970	1971	1972	1973
Nuclear capacity (Mw)	1,027	7,498	10,041	14,828	20,105
% of electricity capacity	0.4	2.2	2.7	3.7	4.6
Gross output ('000 Mwh)	4,368	23,750	40,724	57,705	88,097
Reactor development costs ($ m.)	192	220	221	290	332
Of which nuclear safety ($ m.)	19	35	34	46	57
AEC appropriations ($ m.)	2,625	2,222	2,308	2,293	2,633
AEC plant investment ($ m.)	8,871	9,728	9,925	10,152	10,489
Employment ('000)	133.9	122.5	115.2	112.9	112.5
Uranium ore reserves ('000 tons of contained U_3O_8)	145	243	275	273	273
Ore milling capacity ('000 tons per day)	12	28	28	32	30

Source: U.S. Bureau of the Census, *Statistical Abstract of the United States, 1974* (Washington, D.C.: U.S. Government Printing Office, 1974), pp. 527, 528.

been the fastest-growing sector. Second, nuclear-produced electricity is still far from competitive on cost grounds: Nordhaus quotes 1970 comparisons of $2.51 per million Btus for light water reactors, $1.62 for electricity produced from natural gas, $0.96 for high quality shale oil, $0.29 for strip-mined coal, and $0.01 for Persian Gulf oil.[6] Despite convergence since then between nuclear-power costs and the alternatives, a convergence likely to continue in the future, the differentials remain wide. Third, the rate of progress of nuclear energy is not solely or even primarily determined by market forces, but is a function of government promotion. The most important variable may be the federal R and D budget for nuclear energy. In the 1974 energy budget nuclear energy accounted for 73 percent of the total, or $563 out of $771 million.

Installed nuclear power plant capacity might fall anywhere within the range 240-450 thousand Mw in the United States by 1985. AEC projections of capacity indicate 131,000 Mw in 1980, 500,000 in 1990, and 1,200,000 in the year 2000; by then nuclear plants might provide 60 percent of US electric power. (World nuclear power forecasts are given in Chapter 1.) The National Petroleum Council came out in favor of a lower figure for the year 2000, in the range 750-950,000 Mw.[7] Forecasts of electric-power demands tend to fall within the 1.0-1.7 million Mw range for the year 2000, with the nuclear proportion between 50 and 60 percent. These estimates imply a much smaller nuclear power industry than predicted by the AEC. At this time the future is too uncertain for precise predictions to be made: rising prices may dampen the growth in total energy demand; the rate of technical change will have a marked influence on the pace of nuclear power; coal gasification and liquefaction may become very competitive. Two predictions are fairly safe: there will be a substantial nuclear-power industry by the end of the century, probably the dominant source of electric power; and there will still be substantial demands for fossil fuels—for electric power, for other energy needs and as chemical feedstock.

Resources

Uranium resources will not present a problem once breeder reactors are adopted, because it will then become feasible to use the abundant U-238 and thorium-232 rather than the relatively scarce U-235. Reserve estimates vary with new discoveries, changes in extraction costs, improvements in nuclear-power efficiency, and subjective interpretation of the evidence, but provided that there are sufficient initial supplies of U-235 at the outset of the breeder era, the resource constraint must disappear, partly because breeding itself creates more fuel than is used, partly because of the use made of the much more abundant U-238.

It follows that the uranium supply situation need be a matter of concern only in the short and medium runs. Even here, it is difficult to appear very worried. First, the search effort for new discoveries has not been very intense relative to the search for oil and gas. Second, to the extent that there is a problem it is not one of absolute resource scarcity but of higher costs associated with increasing supplies. Costs of $100 per pound are not practicable until most of the ore can be used, i.e. until breeders are adopted. However, even before then, the cost of U_3O_8 is such a small proportion of the cost of nuclear electricity that substantially higher costs could be absorbed without making nuclear power prohibitively expensive. According to Lovins, quoting an AEC commissioner, an increase of $1 per pound in U_3O_8 would raise the production cost of bulk electricity in a 1-Gw PWR by only $0.000052 per kwh without, or $0.000034 per kwh with, Pu recycle.[8] Put another way, a tenfold increase in U_3O_8 —enough to raise US reserves by about 2000 percent without environmental hazard— would raise nuclear electricity prices by only 18-27 percent, hardly staggering compared with recent increases in energy costs.

Because of rising costs and changes in estimates of reserves due to new discoveries and revisions in the volume of existing resources, the uranium resource situation is subject to continuous revaluation. In January 1974 the AEC suggested the following estimates at varying price levels (in thousands of tons): 277 $8; 340 $10; 520 $15; 700 $30. These figures refer to known reserves. The equivalent potential reserves were: 450; 510; 570; 1,180.[9] Some perspective is given to these estimates if they are compared with total US production between 1948 and 1973 of 263,000 tons and estimated requirements for the period 1973-2000 of 1.8 million tons. Despite potentially large demands for nuclear fuels in Western Europe, imports are another possible source. Noncommunist world reserves (including the US) are now estimated at 1240×10^3 tons at $10 per pound and 2160×10^3 at $15 per pound. Potential reserves at these price levels are approximately double these levels. The higher prices induce new uranium suppliers into the economically feasible range: for instance, at $15 per pound Sweden could supply 350,000 tons of known reserves.

Although each new estimate of reserve availability at given price levels tends to be a little more pessimistic than the one replaced—largely because of rising mining and milling costs—the uranium supply position will not be critical unless resources were exhausted before the arrival of breeder reactors. Once most of the ore is usable, even extravagantly high uranium prices may be consistent with feasible power production. At $100 per pound it becomes feasible to extract uranium from shale, at $200 from granite, and at $500 from seawater. Such prices could hardly be justified under the predictions of any convincing scenario about future energy prices, but they illustrate the familiar theme that *physical* resource availability is not the problem.

As far as nuclear fuels are concerned, AEC's main worry relates not to ore reserves nor to their price but to possible bottlenecks in production—both in respect to mine and mill complexes and to enrichment plants (used to separate the uranium into a high U-235 content product stream and a lower U-235 content "waste" stream). Without imports of uranium twelve to fourteen new mill complexes would be needed by 1980; assuming maximum imports six or seven complexes would still be required. Lower grade ores can be utilized only if investment decisions in exploration and milling are made at the right time. Drilling rates would need to be three to four times current levels if nuclear power forecasts are to be met. Demand up to the year 2000 will be met largely from new rather than existing mills (perhaps 90 percent of the total).

Although the three existing enrichment plants (at Oak Ridge, Tennessee, Paducah, Kentucky, and Portsmouth, Ohio) will be able to meet a much higher proportion of demand for separative work, the bottlenecks in enrichment plant could be serious. The capital cost of an 8.75-million-SWU-per-year plant would be very heavy (about $1.5 billion), and there is technological debate about the relative value of two different processes—the gaseous diffusion and the gas centrifuge methods. The lead-times for both mills and enrichment plants are very long—about eight years. Delays in installing enrichment plants would require more fuel and hence more mills. Additional capacity in milling could be created only if the decisions were taken well in advance. There is insufficient coordination between the different parts of the nuclear-power industry. Neither the AEC nor the private firms involved in the nuclear fuel cycle have taken the required investment decisions in regard to enrichment, while independent utility companies have committed themselves to nuclear-power plants without any attempt to explore the implications of total demand for nuclear fuels. Yet in relation to *total* investment requirements for the industry as a whole, investment in enrichment plants is, despite its large absolute size, quite modest: an estimated $9 billion up to the year 2000 compared with $18 billion in exploration, mining, and milling and over $500 billion in nuclear-power plants. Assuming that the breeder would be introduced by 1986 and would penetrate the market at the same rate as LWRs after 1967, the AEC estimated that six new separation plants of 8.75 million SWU capacity would be needed by the year 2000, and if the breeder adoption date was deferred until 1990 fuel requirements might increase by 20 percent. Furthermore, if nuclear-power plants of the burner type continue to be introduced at the present rate by utility companies, bottlenecks in uranium production capacity might develop by the mid-1980s. It is doubtful whether measures taken so far provide sufficient protection against these contingencies.

Health Hazards

There are two overriding considerations in any discussion of the health risks from nuclear-power production. First, a distinction must be drawn between health hazards to the operators of nuclear-power plants, uranium mines, and other industrial operations connected with nuclear power and those affecting the general population at large. Second, the appropriate criterion for comparison is not the ideal but unattainable absolute standard of zero risk but the relative standard of the health hazards associated with fossil-fuel consumption.

Despite intermittent accusations blaming radiation exposure for subsequent leukemia deaths among nuclear-power-plant workers in several countries, the official US statistics are very encouraging about the occupational accident and disease record of the nuclear-energy industry. Between 1943 and 1970 the accident frequency rate among AEC workers was consistently 25-50 percent of the average for all industries. Of 295 deaths over the same period in the AEC program, three-fifths were construction accidents and only six or seven due to radiation exposure. The incidence of mortality among AEC workers at Oak Ridge has been only about 70 percent of what might have been expected on the basis of age-adjusted mortality rates, though socioeconomic influences might help to explain this. Between 1947 and 1970 less than 0.2 percent of workers were exposed to an annual dose of more than 5 rem, the maximum permissible level. Uranium mines have had a much poorer record, partly due to lax state safety standards in the past. Nevertheless, "occupational health effects from accidents and chronic disease are substantially greater for coal mining than for uranium mining and milling per megawatt of power generated."[10]

The evidence on "normal" exposure from nuclear-power plants as it affects the surrounding population is even more reassuring. Exposures of gases, halogens and particulates, and mixed fission products are in all cases very much lower than the permissible releases allowed in the licenses to the reactor operators. More detailed studies of individual installations reinforce this generalization. For instance, the Indian Point PWR on the Hudson River results in contamination of fish. But the doses are so low that the annual dose due to a steady diet of Hudson River fish would be equivalent to 120 hours per year of exposure to the natural igneous rock of Manhattan Island. The main worries, however, relate not to the present but to the future, when nuclear energy is more widespread than today. But, on the assumption of 500,000 Mw by the year 2000, the dose might increase from the present 0.002 mrem per year to 0.4 mrem. This startling increase fades into relative insignificance when compared with the 4 mrems per year from weapons fallout, 72 mrems from diagnostic use of medical X-rays and radioisotopes, and 130 mrems from natural background sources. Put another way, exposure from the nuclear-power industry might produce a maximum of 12 cases of cancer per year in the US population by the year 2000 compared with about 500,000 cancers from other causes.

The comparison with the health effects of fossil fuels generation is also impressive. With regard to air pollution, a nuclear PWR offers about 18,000 times less health risk than a coal-burning power plant, while a BWR with a thirty-minute hold-up of stack gases offers perhaps twenty-four times less health risk. Although these findings may be suggestive rather than precise, and more research needs to be done, e.g. on the toxic, mutagenic, and teratogenic effects of radionuclides in low concentrations, the case for the view that conversion to nuclear fuel would mean a healthier population and fewer occupational health risks than with fossil fuels is overwhelmingly strong. Of course, these comparisons do not shed light on other dangers associated with nuclear power: the problems of radioactive waste disposal, the risks of a major disaster at a nuclear-power plant, or the possibility of sabotage, political blackmail, or similar strategic objections.

Disposal of Radioactive Wastes

The disposal of radioactive wastes is a consequential problem of the use of nuclear energy that is, quite rightly, arousing anxiety. These anxieties refer not so much to the volumes currently stored as to the accumulations in the long run as nuclear power production expands. An AEC Advisory Committee set up by the National Academy of Sciences, National Research Council laid down three basic principles to govern radioactive waste disposal:

1. All radioactive wastes are biologically harmful, and should be isolated from the biological environment over their period of harmfulness, in some cases more than 600 years.[d]

2. Waste generation is more or less proportional to nuclear power production, and such production is near the bottom of a steep exponential curve. Accordingly, criteria for waste disposal practice should be geared to later levels of production rather than to what is safe at the current low levels.

3. Safety should not be sacrificed to cost economies in waste disposal.

Research and development is under way into more effective and safer storage methods, but at present it is probable that the above principles are not being met. High-level aqueous-solid slurries can be reduced to solids and stored either in underground natural salt beds impervious to ground-water flow or in concrete-and-metal bins. This is not too bad a solution. The disposal of intermediate- and low-level wastes is much less satisfactory, involving dubious procedures such as burying in shale or earth ponds, direct discharge into the sea, the use of shallow trenches, or discharge of gases into the atmosphere from tall dispersion stacks.

On the credit side, disposal costs are still small (say less than 2 percent of production costs), and the industry could afford more expensive disposal methods if required. It is arguable that control of waste disposal should be

---

[d]The AEC laid down as a practical rule of thumb that 20 half-lives should be regarded as the minimum period before high-level radioactive wastes should be considered safe for biological exposure. A half-life is the time required for a radioactive material to decay to one-half of its original mass. Long-lived isotopes such as strontium-90 and cesium-137 have half-lives of 28-30 years. The artificial isotope Pu-239 has a half-life of 24,400 years. Since world inventories of Pu-239 may reach thousands of tons in the 1980s, while a lethal dose for the world population might be contained in a piece the size of an orange, its disposal problems appear immense.

placed in the hands of a committed "watchdog" agency independent of both the nuclear power industry and government.

Another solution to the disposal of very long-life high-level wastes that has received recent research attention without yet solving all the practical problems is to irradiate, or burn up, the wastes in the reactor. These wastes, called *actinides*—the two most common being americium and curium—could in this way have their active lives reduced to less than 500 years, putting them in the same category as other fission products from nuclear reactors. The theoretical aspects of this solution are well known. Two practical problems on which research is under way are how to separate the actinides efficiently from other wastes and how to design a fuel can tough enough to contain them while being burnt up in the reactor.

Nuclear Theft

One of the risks associated with a heavy reliance on nuclear power is the danger that unscrupulous individuals or groups may steal fissile materials for the purpose of making nuclear weapons to be used either as political blackmail or for monetary gains.[11] Nuclear weapons are not difficult to make given the availability of the required raw materials (plutonium, high-enriched uranium or U-233). Even by 1980 scores of thousands of fissile materials will be present in the world's nuclear-power industry, and opportunities for theft will arise in enrichment plants, transportation, or storage. Of course, safeguards have been introduced by the AEC in the United States and by similar bodies in other countries. Nevertheless, safeguards remain inadequate, particularly in respect to relatively small amounts (e.g. less than two kilograms of plutonium) and in transit. Technological solutions to safeguard problems are relatively easy to devise. The institutional requirements are trickier, but are soluble. The main needs are: a system of safeguards for each fuel cycle; the establishment of a federal nuclear materials security service; a set of procedures for evaluation of the security plans submitted by industry licensees; and development of a common international policy. The justification for serious attention to these questions is that hitherto "in our exploitation of atoms for peace, we have been concerned primarily with the dangers of a malfunctioning machine. We can no longer ignore the dangers of a malfunctioning man."[12]

Conclusions

The prospects for nuclear power may be over-optimistic, for several reasons. First, although the safety evidence is convincing on the basis of published data, there are some doubts, because an intermittent, if not consistent, tendency to suppress unfavorable research results has been noticed in the AEC.[13] Sometimes this body has appeared to be more concerned about public relations than objective safety standards. One possible result of the recent separation of research from safety and regulation by dividing the AEC's responsibilities between the Energy Research and Development Administration (ERDA) and the Nuclear Regulatory Commission (NRC) is a reduction in the conflict of interest that has handicapped frank assessment of the teething troubles of the nuclear-power industry. Second, the disposal of radioactive wastes problem is a big unknown with uncertain repercussions. It is possible that the frightening effects of error may lead to an overestimation of the risks, but the technical and human elements in the situation are complex if not intractable. Third, the other environmental problems of a nuclear-based economy are far from negligible (see Chapter 10).

Moreover, the implications of relative price changes and bottlenecks may have been misread in ways favorable to nuclear power. The rising price of uranium, particularly because of cost considerations in the extraction of new reserves, may switch the economics of power production heavily in favor of coal in the medium term before breeders are introduced. The ability to import uranium at moderate cost is likely to be severely constrained by supply limitations, competing demands from other countries—especially in Europe—and by nationalistic resource-conservationist policies pursued by uranium-producing countries. Perhaps even more serious, there are capacity constraints in the separation process stage of production that are hard to eliminate in view of the current levels and dynamics of capital costs. It is possible that the nuclear power plants already scheduled to come into production in the late 1970s and early 1980s may lack the fuel to operate.

On the other hand, in the absence of complete certainty and perfect futures markets there is a danger that a sequence of short-run decisions taken on the basis of current relative price changes may result in long-term commitments that are both uneconomic and inflexible in the future. Long lead-times, the interactions between competing fuel prospects and R and D decisions, possible conflict between government interests and those of the utility companies, all exacerbate this fact. If there were some consensus that nuclear power was the backstop technology and a belief that the technical difficulties in safety of operation, fuel supply, efficient production, and waste disposal were soluble—as yet, this consensus does not exist, a commitment to a long-term energy policy with a heavy reliance on nuclear power would hasten the resolution of technical problems and permit wiser short-run decisions in respect to other fuels.

7

Other Fuels

Shale Oil

Shale oil is an old energy resource; there was a thriving industry in Scotland from the early 1850s until it was eliminated by competition from cheaper fuels about a century later. The rising fuel prices of recent years make the prospect of a revival of the shale oil industry, this time in the United States, a real possibility. Oil shale is a rock containing a tarlike substance called *kerogen*, which undergoes pyrolysis to yield raw shale oil. This may be refined into petroleum products via standard refinery techniques. The major deposits in the United States are located in the Green River Formation of the Piceance Basin of northwest Colorado, the Uinta Basin of eastern Utah, and the Green River Basin of southwest Wyoming. These fields differ greatly in quality of grade, and this is important because commercial production requires yields in excess of 25 gallons per ton of processed shale. Only about one-third of the known reserves are above this quality, and the richest reserves are within an area of 3000 square miles in Colorado. It is this area that has been the main focus of the revived interest. Reserve estimates have been revised from time to time, usually in an upward direction, but the most recent estimates indicate at least 1800 billion barrels of oil in the main areas mentioned above. Most of the resources are federally owned (at least 80 percent of the reserves), and a new oil leasing program was introduced in 1971 after forty years of inaction.

It is difficult to evaluate the commercial profitability of oil-shale production, particularly because of rapidly escalating capital costs. For instance, ARCO postponed a large shale-oil project in Colorado when capital costs almost doubled. A major uncertainty is, of course, the price of oil, though if oil prices continued at their present levels it is probable that shale-oil extraction would be viable. Demonstration projects for above-ground retorting systems have been under way since the mid-1960s. The prospects for large-scale oil-shale production depend, in part, upon the interested companies' assessment of the risks and uncertainties.

However, there are other serious constraints related to the impact of shale-oil production on the environment. First, it is possible that surface mining of thick deposits may become the most profitable type of mining. Second, since shale oil is much too heavy to be transported, and processing uses substantial volumes of water, available water supplies may be a serious problem limiting production. Most important of all, production levels in excess of 200,000 barrels a day from shale plants will probably infringe the 1980 Colorado SO_2 standards of 10 mg./m^3. There are other problems too: land-use disturbance, a large volume of solid waste (more than the original volume extracted), increased salinity of the Colorado River, disturbance to vegetation, wildlife and stock watering, and the problems of population concentration in a sparsely populated region. These constraints may inhibit oil-shale development even if economic conditions are favorable. PIB estimated a constrained production level of 250,000 barrels per day by 1985. If the environmental problems could be solved, if the government leased more lands, and if high oil prices were maintained, it is possible that production could reach 1 MBD by 1985. The coincidence of these stringent conditions is unlikely. In any event, the contribution of shale oil to the nation's energy needs will remain either small or negligible.

Synthetic Fuels

In view of the relative abundance of coal, and to a lesser extent of oil shale, in the United States, and the identification of the energy problem as a petroleum, and perhaps a natural gas, shortage, many difficulties could be solved if a magic wand could be waved to transform the plentiful into the scarce fuels. In fact, this can be done, though only with difficulty. Of course, town gas has been produced in both the United States and in some European countries from coal since the nineteenth century. In Germany between the wars several processes were developed for gasifying coal, of which the best known is the Lurgi process, and plants were built to supply Germany's wartime fuel needs. There was also some synthetics research in the United States in the 1940s and early 1950s, but projects were abandoned because the fuels would have been uncompetitive. Interest has now revived again, obviously in response to the higher fuel prices and fears of depletion of domestic oil and gas supplies, but also because of the possibility of new technologies that might cut production costs by 25 percent and raise thermal efficiencies from 60 percent (under the old Lurgi process) to about 75 percent.

Substitute natural gas (SNG), synthetic crude (syncrude), and methyl alcohol (methanol) can all be produced from coal or from oil shale. Most of the new processes have reached the developmental and, in a few cases, the pilot stage, usually with a degree of government sponsorship. However, there is insufficient experience available on which to form a judgment about the most likely technology, and even less about the prospects for commercial production. Much depends on government policy, and in particular upon whether incentives would be provided to stimulate synthetics research and production. Market acceptance for the products is unlikely to present difficulties, since they are, with one or two exceptions, easily substitutable for existing natural fuels. The environmental constraints are quite severe, however. Low operating costs demand very large-scale plants, water requirements are heavy, and waste runoff can be an environmental hazard.

The MIT study[1] made some estimates of capital and operating costs for plants of 250 million cubic feet of SNG per day (or equivalent output of syncrude or methanol) from coal and (except for methanol) from shale. The data are already obsolete because of rising costs (the estimates are for 1973), but there is little reason to expect the relative costs to vary much. It is also notoriously difficult to predict capital costs for technologies not in commercial production. Nevertheless, the capital costs estimates were very similar for all fuels and technologies, converging around $350-400 million. Even operating costs clustered in the range of $1.20 to $1.60 per million Btu, though shale-oil costs for syncrude were a little lower. Synthetic fuel production is highly capital intensive. Capital-sales ratios would probably be about 3:1, compared to 1:1 in other capital-intensive industries such as chemicals. This is not necessarily a problem, but it does raise difficulties. The volume of capital at risk in a single plant is huge. If technological risks were low and if the stability (or increase) in unit sale prices was assured, this would be all right. Unfortunately, the current state of synthetics technology and the uncertainties of the energy situation do not guarantee these preconditions.

The prospects for synthetic fuels making major inroads into the fuel supply position in the near future are poor. There are many reasons to support this view. The technological uncertainties have not yet been resolved. The federal government has yet to formulate a clear policy on synthetic fuels, covering such important questions as the scale of research, the degree of public involvement, the provision of incentives, and whether environmental constraints can be relaxed. The capital risks are likely to deter major projects until some of the uncertainties have been reduced. It would take a great many plants to meet a high proportion of supply. One hundred of the 250 million cubic feet per day plants of SNG would be required to meet one third of the country's gas needs, and would consume the total current output of coal. One hundred syncrude plants would be needed to meet one quarter of the country's petroleum requirements. The lead-times are very long, in view of the need for pilot projects, the long construction period, and the permits and environmental objection delays. The PIB forecast no synthetic fuel production by 1985 unless special promotion policies were adopted (the accelerated supply case). Even under these conditions SNG would equal only 1 TCF per year (3 percent of gas consumption) and syncrude only 0.5 MBD (2-3 percent of petroleum consumption). The real breakthrough would not take place until the end of the 1980s or in the 1990s. The competitiveness of SNG with natural gas is likely to depend on the latter's availability rather than relative costs, because synthetic high-Btu gas costs more than $2 per million Btu to produce. Syncrude is perhaps more competitive, since it would be cheaper than $11 per barrel oil if coal costs less than $19 per ton.

Solar Power

In view of the dependence of the world economy on exhaustible energy resources, particularly fossil fuels, there is considerable appeal in development of practical methods of using renewable resources on a substantial scale. Solar energy is an obvious candidate for this role. Not only is solar radiation the energy source, via photosynthesis, for maintaining the biological system as a whole, but it has been used since the beginning of time for more direct energy uses (e.g. drying crops, solar-related wind power) than supplying food and animals for transport. The attractions of solar power are immense. It is very abundant; the thermal power intercepted by the earth is about 17.7×10^{16} watts, many scores of thousands greater than the world's electric-power capacity. Even allowing for reflection and scatter by clouds and dust, the quantity of energy falling on the United States is 700 times the total current energy demand. It is inexhaustible, at least in relation to the future lifespan of the human race. It is fairly universal, though its early applications have been, and will probably continue to be, in sunny regions. Its raw cost is very low, though the costs of collection and storage are high, especially for large-scale uses. Finally, it is environmentally clean.

These advantages are offset by two serious drawbacks that are obvious. First, it is very diffuse (about 0.139 watts/cm^2), and hence a large area is needed to collect large amounts of energy. This explains the suggestions for obtaining the power for solar-power plants by collector surfaces taking up large tracts of desert, such as in Arizona. Thus, using solar power on an extensive scale raises severe land-use problems, or restricts its application to a few special regions accessible to large areas of unusable land and exposed to considerable sunlight. Moreover, the costs of collector surfaces—again, particularly for large-scale projects—may be very high. Another problem arising from the diffuseness is that it becomes very costly to generate high temperatures. This explains why one of the most promising applications is for background heating, hot-water supply, and air-conditioning in homes and commercial buildings rather than as the sole power source in these cases. The second problem is its variability according to time of day, season, and general climatic conditions. The solution to this obstacle is storage, but this raises costs.

High collection and storage costs are the major constraint on the widespread application of solar power. Although some important technical difficulties need to be resolved, and in view of the limited research in this area hitherto there is major scope for technological improvements, the feasibility of solar technology is proven—for conversion to both low- and high-temperature heat, to drive electrical and mechanical motors, and for materials separation. The rate of application of solar energy is a question of economics, not technology. Since it can be employed for almost all known energy uses, the extent of its use depends upon competitive costs. Because it is an environmentally attractive energy source, solar power would, of course, make more headway if fuels were priced at their real social costs, including full allowance for environmental damages. However, even in terms of private cost accounting it is becoming competitive with other fuels for certain restricted uses. Because of the negligible raw-energy cost, the critical determinant is the movement of solar conversion costs relative to the costs of extracting, producing, and delivering other fuels.

There are several ways of using solar energy. One of the simplest, appropriate for heating and cooling buildings, is to heat a liquid enclosed in an exposed collector surface which can either be used directly or can transfer heat to storage equipment or to heat-operated refrigerators. Higher temperatures can be obtained by concentrating solar radiation in a parabolic-contoured "dish" which can focus diffuse heat on a small heat-absorbing transfer surface. Demonstration experiments have been carried out for operating small engines and for high temperature furnaces. Many applications involve several phases—converting solar energy to heat, then to mechanical power or to electricity generation, for example—some of them natural, wind power, or ocean thermal conversion using surface waters to drive turbines, or bioconversion of plants and solid wastes into energy. One of the more interesting methods is direct generation of electricity from solar energy (photovoltaic conversion), making use of semiconducting materials such as silicon and cadmium sulfide. These have been used on space missions, the only major application hitherto. Although 15 percent of the sun's radiation can be converted into electricity, the cost of the materials and of fabrication makes them a thousand times more costly than conventional electric generating plants, and hence impractical without some major technological breakthrough. Although the costs of generating electricity by thermal processes are cheaper than by the photovoltaic method, they remain many times more expensive than by conventional methods. The diffuseness problem is intractable. Either a concentrating device is required (lens, focusing mirror, solar tracking mechanism) or the low efficiency of engines requires large collection areas. The technologies are well known, but the economics make solar energy as a source for large-scale power generation impractical, until major cost reductions are achieved via R and D efforts.

This suggests that solar power for heating and cooling buildings via rooftop collectors will be the major use in the short run. Space heating, water heating, and cooling can be supplied by solar heat at costs less than double those of oil and gas, and in sunnier regions at a cost lower than by electricity.[2] Probably, these systems will continue to be used in combination with conventional fuels for cooking and other domestic uses. A major constraint on the adoption rate is that it is inefficient to convert existing houses to these systems. Thus, market penetration is limited by the rate of growth of the housing stock. A second short-term application is the use of wind energy systems to produce cheap electricity (up to 1950 there were about 50,000 windmills in use for this purpose in the United States). The PIB anticipates a rapid rise in windmills in the 1980s, on the assumption of a halving of capital costs per Kw by 1985.

These considerations suggest that solar energy will make only a marginal contribution to satisfying energy requirements in the near future. The major field of application is for certain restricted purposes in small, dispersed units (in effect, in individual buildings). The Ford Foundation study suggested that savings might vary between 0.3 and 1.0 Q. by the year 2000, the latter target being dependent on public promotion resulting in one-third of new construction involving solar energy units. Given a major federal R and D effort and other policies favoring solar energy, it guessed that one Q. of electricity might be produced by central solar generating stations by 2000. The PIB made no forecast of this type of application, but suggested a range of 0.3-0.6 Q. for space heating by 1985, according to whether incentives were offered. The assumptions behind these forecasts were based on the hypothesis that market penetration is achieved when solar heating costs less than other fuels. The estimates may underrate the extent of consumer resistance or the time lags involved in the adoption and spread of new energy systems.

Geothermal Energy

There are a few geothermal fields that are dry-steam or vapor reservoirs. These include the geysers of California, Laradello, Italy, and New Zealand. In these cases electricity can be produced easily by drilling a hole and piping the natural steam to a nearby turbine. Fields of this kind are rare, however, and the alternatives, hot water or other liquids such as brine and hot rocks, though plentiful, raise much more serious technical problems. For instance, the large reservoir under Imperial Valley in California consists of highly corrosive brines that can clog up drill holes and destroy machinery. The problem of scaling in the nozzle might be dealt with by using polymers for the piping and titanium metal coated with corrosion-resistant titanium nitride for the turbine wheels. The latter is very expensive, though proponents of this source believe that geothermal electricity from this source would be fully competitive in terms of price with coal-, oil-, or nuclear-produced electricity.

These are early days for geothermal technology; it is difficult to evaluate how important this source might become before the end of the century. Most of the technical problems are soluble, however, and if costs, resource constraints, or technological failures in other sectors (e.g. nuclear power) become serious there is little doubt that geothermal power could be used extensively. The combination of circumstances that would create these conditions is unlikely. One official committee predicted 132,000 Mw of capacity by 1985, but the NPC prediction of 3500 Mw may be much closer to the mark. Major technical breakthroughs are needed to pierce the marginal contribution barrier. These include the use of binary fluid heat-exchange methods to exploit relatively low temperature resources; cost-sharing via the development of multipurpose projects for desalination and by-product chemicals production; new and cheaper drilling methods; improved rock-fracturing techniques to increase permeability; the application of geothermal energy for nonelectrical purposes such as space heating; and overall cost reductions to make geothermal energy competitive with other fuels. Also, there are several environmental hazards associated with geothermal power: the disposal of waste water which contains high concentration of salts, boron, ammonia, arsenic and other compounds; thermal, noise, and air pollution (H_2S); subsidence over hot-water reservoirs; and the risk of earthquakes due to reinjection.[3]

Despite the difficulties, geothermal resources are plentiful. The U.S. Geological Survey estimates that the heat in the top ten miles of the earth's crust amounts to about 3×10^{26} calories, or 2000 times the heat generated by the world's total supply of coal. Of course, the dispersion and depth of this heat rules much of it out for commercial exploitation. Also, current technology converts only about one percent of the stored energy of hot-water reservoirs into electrical energy. However, if one-tenth of the geothermal energy in the top two miles of the earth's crust could be extracted, this would supply 58,000 Mw each year for more than a half century. One industry executive believes that under Imperial Valley there is enough energy to meet the demand for electricity in the Southwest for 200 years.

There are three ways to produce geothermal electricity from hot-water sources: the flashed-steam process, the binary vapor technique, and the more efficient total-flow system, using liquid and steam together. The first is used in Cerro Prieto, Mexico; the second will be used by the Southern Californian Edison Company beneath Mammoth, California; and the third is still in the experimental stage. Even larger resources are available in the form of hot dry-rock formations, usually granite. One estimate is that in the thirteen western states there are 95,000 square miles which have dry rock of 550° F. at a depth of 3.5 miles. One possibility is to pump water down to the rock to create steam in a manner similar to the way of nature. The problem here is devising cheaper and more effective drilling techniques. In many other parts of the country there are hot rock deposits where the temperature is too low to generate electricity but where possibilities exist for home heating, hot water, air-conditioning, and similar uses. There are experimental power plant projects planned in Montana and Nevada as well as an extension to the geysers dry-steam field north of San Francisco. In addition, many private geothermal developers have leased land from the federal government. Although geothermal power is unlikely to have more than a negligible role in meeting energy demands over the next decade or two, it is more practical and feasible than the more exotic forms of power such as hydrogen, solar, and even tidal power.

Tidal Power

Using the tides as a source of energy has been considered intermittently for decades. Feasibility studies were undertaken in the 1930s to see if the tides in Passamaquoddy Bay, Maine could be used for power. Due to lack of resources, realization that the costs would involve unit electricity costs far higher than prevailing prices using other fuels, and high transmission costs (not now a serious problem), the project was never completed. The obvious physical requirement for tidal power is a large tidal amplitude through geomorphic rocks requiring a relatively small dam to harness the power. This can be obtained at many places along the world's coasts, though rarely close to major consuming centers. Although rising energy prices make the capital costs of tidal-power plants less prohibitive than in the past, this remains a source unlikely to be exploited on any scale in the foreseeable future. Related alternatives such as making use of surface waves or the ocean's internal waves are even less likely, though no doubt technically feasible.[4]

The existence of a tidal-power plant in Brittany is evidence that the obstacles to the use of tidal power are economic rather than technical. There is a big difference between high and low tide of 13.5 meters in the Rance Estuary on the English Channel, and a maximum flow of 18,000 square meters per second. The estuary is 750 meters wide at the point where the tidal-power plant is located. The project was begun in 1961 and the power plant came into operation in 1966. It produces some 608,000 kilowatts, of which almost 11 percent are used for pumping.

Hydrogen

In view of the growing scarcity of fossil fuels, the severe attack of hiccoughs in the progress of nuclear power, and the economic obstacles in the way of making major use of solar power, not to mention the locational constraints on the extent of application of geothermal and tidal power, the search for an alternative fuel goes on. One possibility that has received some attention is the use of synthetic hydrogen as a universal fuel. Out of several theoretical possibilities that conform to the criteria of relative abundance and environmental cleanliness, hydrogen appears one of the best bets on grounds of cheapness—though still many times more expensive than fossil fuels. Potentially, it could satisfy all the varied demands for energy—industrial, commercial, residential, and transportation.

In the current state of technology, the most obvious method for producing hydrogen synthetically is via the electrolysis of water. This method is a high consumer of electricity, and high production costs could be redeemed only partially by low transmission and distribution costs and by the sale of by-products such as oxygen and heavy water. The conversion of low-grade fossil fuels to hydrogen is another possibility, while splitting water into hydrogen and oxygen by using the heat from either a nuclear reactor or a solar furnace is still only at the research stage.

Transmission problems are unlikely to be serious. Existing pipelines (50 miles long near Houston, a system of 130 miles in West Germany) are relatively short, and compressors will be needed for long-distance transmission, but the difficulties are hardly intractable. Estimates of hydrogen transmission costs suggest that they will be only 20-33 percent of overhead electricity transmission lines and about one percent of the cost of underground electric power cables—measured in terms of million Btu/100 miles. The storage problem can be handled either by storing in underground porous rock formations or by liquefaction in refrigerated tanks.

There is no reason why hydrogen cannot be used for all the purposes served by natural gas today. Moreover, by the technique of catalytic combustion it may be feasible to make hydrogen a completely clean fuel by eliminating even trace nitrogen oxide emissions. The vehicle-use problem is primarily one of tankage— to make tanks light and small enough. One solution might be to use chemically bound hydrogen in metal hydrides which can be decomposed by the heat of the exhaust to provide pure hydrogen. Other potential uses include aircraft, industry (chemicals, fertilizers, and iron production), and fuel cells.

There is a safety problem because of its high inflammability range, very low ignition energy (easily ignited by a static spark) and very high flame speed. However, old town gas contained 50 percent hydrogen and was also toxic because of its carbon monoxide content. Also, hydrogen can easily be detected because of odor, and because of its lightness diffuses away quickly from a leak or spill. Most important of all, there is considerable practical experience in handling in industry and in the aerospace sector.

Even if all the technical problems were solved, hydrogen is no panacea. It is not a primary source of energy, requiring energy inputs to produce it. However, it has several advantages—flexibility, cheap transmission, and environmental purity—and it will be an attractive fuel, particularly when produced via the use of nonfossil energy sources. But it is unlikely to make major inroads over the next two or three decades.[5]

Part III:
Problems and Solutions

# 8

OPEC, The United States, and World Oil

Production, Consumption, and Reserves

As everyone knows, the dominating feature of world petroleum production is the preeminence of the Middle East, accounting for over one-third of world production and 55 percent of world reserves (Table 8-1). If North Africa is added, the shares become 43 and 63 percent respectively (production and reserves). Taking in the remaining OPEC countries raises these proportions to one-half and seven-tenths. Since all the oil producers in this group are negligible consumers compared with such areas as North America, Western Europe, and Japan, these data go far in explaining why market forces do not operate effectively in the world oil market—though surprisingly they have had considerable impact until very recently. The absence of Western Europe (with the exception of the nations with access to the relatively limited supplies of North Sea oil) and Japan among producers means that these areas with oil-dependent technology have far less freedom of maneuver than the United States. The US still produces 20 percent of the world's oil and, although she possesses little more than 5 percent of proven reserves, her position would be very healthy, were it not for very high energy consumption and the relatively inelastic demand for gasoline.

The dominance of the Middle East in world oil production is quite recent. Before World War II three countries (the United States, the USSR, and Venezuela) accounted for more than 80 percent of world output, compared to the Middle East share of less than 6 percent. Since then, a stream of new discoveries of very low-cost oil combined with escalating demands from the developed countries have steadily boosted the Middle Eastern share to its present levels. Demand has been doubling approximately every decade. This exponential growth cuts heavily into available reserves. Even today's world output equals about 3 percent of known reserves—thirty-five years of production at current rates. But demand will continue to grow. Extrapolating recent growth rates, consumption in the year 2000 could be equal to about 23 percent of current proved reserves. Obviously, world petroleum reserves will be exhausted before then unless either new discoveries are made on a substantial scale or demand growth is moderated. The most likely outcome is a mix of both, with the oil producers themselves controlling demand by fixing prices or output.

There are moderate grounds for optimism in regard to new discoveries. The sharp upward movement in prices will stimulate new exploration. Even before

Table 8-1.
Crude Petroleum Production, Consumption, and Reserves by Major Producing Region, 1973

	Production (MBD)	Consumption (MBD)	Reserves (billion barrels)
Middle East	21.3	1.1	316.0
Communist countries	9.7	8.7	64.4
Africa	5.7	0.9	52.5
United States	10.9	17.2	35.3
Other W. Hemisphere	3.7	4.4	22.0
Non-Communist Europe	0.4	14.5	17.0
Caribbean	3.7	0.5	15.9
Japan	0.02	5.2	0.03
Other E. Hemisphere	2.2	2.7	18.3

Source: Federal Energy Administration, *Project Independence Blueprint* (Washington, D.C.: U.S. Government Printing Office, 1974), pp. 351, 352.

prices rose, however, there have been increases in the stock of reserves. Major new producers such as Nigeria have entered the market. Although discoveries in the North Sea and Alaska have made only marginal additions to world reserves, the Alaskan fields have hardly been touched. Discoveries offshore in such areas as the Gulf of Mexico, the North Atlantic, the Pacific, and more recently in the Persian Gulf raise the hope of more offshore fields in these and other areas. Also, there have been recent small-scale discoveries in some Latin American countries (e.g. Bolivia) so that it is at least a possibility that other Latin American countries may join Venezuela, Mexico, and Ecuador as major producers. Nevertheless, there is little evidence in recent discovery experience that the Middle Eastern-North African hold on the oil market can be weakened, or that the long-run supply constraints can be avoided. (Saudi Arabia holds 38 percent of the Middle East's reserves, followed by Kuwait with 18 percent and Iran with 17 percent. Libya owns more than one-half and Algeria one-quarter of North African reserves. Apart from the USSR (second only to Saudi Arabia) and the United States, the other leading oil nations (from a reserves point of view) outside the Middle East and North Africa are Nigeria, Venezuela, Indonesia, and Canada.) OPEC's export capacity is expected to increase from 30 MBD in 1973 to more than 53 MBD by 1985. Of this projected total Saudi Arabia alone would account for 36 percent, with Libya and Iran a long way behind with less than 15 percent each. Whether or not this capacity will be reached depends on a myriad of unforeseeable factors: the success of demand restraint in the industrial world; the rate of economic development in developing countries; production decisions within the individual OPEC countries; and the unity and durability of OPEC itself. Assuming that OPEC survives, much depends on the policy pursued by Saudi Arabia.

A Brief History of OPEC

In the 1950s the dominant pattern of concession agreements between the oil companies and the oil-producing countries was that the companies paid the host government a tax on profits realized from the sale of exported oil. The tax rate was 50 percent of profits calculated on a public price—the posted price. The host governments had a strong interest in keeping these posted prices as high as possible, because this boosted their tax receipts. However, rapid expansion in supply associated with new discoveries from the middle 1950s subject the prevailing posted prices to increasing pressure. A gap emerged between the posted price and the market price, oil company tax payments were related to the posted price, and as a result the companies made several reductions in the posted price between 1958 and 1960. The consequent loss in tax receipts and the threat of further losses led to the formation of the Organization of Petroleum Exporting Countries (OPEC) in 1960. Its original members were Iran, Iraq, Kuwait, Saudi Arabia, and Venezuela, but a trickle of new members over the years has brought its present strength up to thirteen.

The primary goal of OPEC in its early years was to keep the posted price as high as possible. It was successful in doing this, so that the posted price in effect became a tax reference price to serve as a basis for tax payments rather than a true market price. Other minor concessions were gained during the 1960s. In 1962 and 1963 an agreement was reached under which royalties were treated as deductions from income rather than as income tax. This change raised the government's share of the take, even though—until they were phased out—the oil companies were allowed partial offsets from the posted price. In 1967 three OPEC members (Venezuela, Libya, and Indonesia) changed over from the old taxing real profits system to the tax reference price system. In 1968 OPEC issued a declaration calling for renegotiation of existing contracts, determination of posted prices by the host governments so that they would move in line with world prices of manufactured goods, and eventual OPEC country participation in concessions.

Up to 1970, however, the consensus is that though OPEC appeared to be winning the minor skirmishes the oil companies were winning the war. As a cartel OPEC was ineffective. The countries were suffering from overproduction in an era when demand was growing more modestly than in more recent years. They differed widely in their reserves, their fiscal needs, and their development potential. They lacked a tradition of cooperation, a mechanism for restricting output, and an identity of interest. The oil companies knew their interests better. They produced, transported, refined, and marketed the oil. They maintained their profits remarkably well and were very successful in their dealings with consumers, so that they could easily afford minor concessions to the host countries.

In 1970-71 everything changed. Quite suddenly, rapid demand growth ran ahead of capacity. There were output cutbacks in Libya, while the United States finally ran out of excess capacity. In September 1970 posted prices made their first notable advance since the formation of OPEC. Crude oil from Libya, Nigeria, and the Mediterranean ports handling Iraqi and Saudi Arabian crude increased in price by about 30 cents per barrel, and the tax rate was raised from 50 to 54-58 percent. Iran and Kuwait followed a similar path in November 1970, and Venezuela in December. These actions were facilitated by the closure of the Trans-Arabian pipeline in May 1970 and the aftermath of the closing of the Suez Canal in the 1967 Arab-Israeli War, which resulted in a shortage of tanker capacity needed to ship Persian Gulf oil via the long route around Africa.

In February 1971 the Tehran Agreement was signed between twenty-two international oil companies and six OPEC members. For the first time the oil companies bargained collectively with the producing countries rather than individually. In effect, there was collusion between the oil-producing countries and the multinational oil companies, despite window-dressing to convey the impression that they were bargaining hard with each other. Moreover, they were encouraged at this stage by the United States. The United States had two main motives: first, to use higher oil revenues for the Arab countries as a "sweetener" to induce them to be more willing to compromise in a political settlement with Israel; second, to allow oil prices to drift upwards appeared to have advantages from the point of view of competition with Western Europe and Japan, since these areas depended more on Middle Eastern supplies but had benefited (despite high fuel taxes) from very cheap oil. The strategy backfired, initially because some of the oil-producing countries quickly recognized that they did not need to cooperate with the oil companies, and subsequently because of the 1973 Arab-Israeli war and the decision to use the supply of oil as a political weapon.

The balance of bargaining power had shifted perceptibly. The agreement included an increase in posted prices of about 33 cents per barrel; upward adjustments in the prices of heavier crude; elimination of discounts; incremental increases in posted prices through 1975; and a minimum tax rate of 55 percent. A month later, a similar agreement was ratified at Tripoli. Prices were in fact higher because of the closer proximity of Mediterranean ports to European markets. Other price supplements included a low-sulfur premium, a Suez Canal closure premium, and a freight rate component.

In the following year the producing countries demanded a higher posted price to take account of the declining purchasing power of the US dollar. Under Geneva I posted prices were raised by 8.5 percent in January 1972. In the following year, a second Geneva agreement (Geneva II) provided for an increase in June of almost 12 percent over the January level to compensate for the second US dollar devaluation. Provision was made for a parity index to adjust prices according to changes in the value of the dollar, and minor price adjustments were made throughout 1973.

A new ingredient was introduced after October 1972 with the signing by several OPEC members of a General Agreement on Participation. The governments would acquire immediately (January 1973) a 25 percent interest in all production facilities, to be raised by fixed increments to 51 percent participation by 1982. Negotiations were also planned for transportation and refinery facilities. To minimize market disruption bridging arrangements were made for the first three years, though oil purchased under these arrangements would be more costly. In April 1973 Nigeria acquired a 35 percent interest in local Shell-British Petroleum properties, with the option to increase the share to 51 percent by 1982 or earlier.

These bitter pills were sugar-sweet compared to what the consuming countries had to swallow after October 1973. With the outbreak of the second Arab-Israeli war, the Arab members of OPEC (with the exception of Iraq) decided to use crude oil as a political weapon by cutting back oil production (by 5 percent per month, though Saudi Arabia announced an immediate 10 percent cutback), imposing an embargo on several countries—particularly the United States and the Netherlands—and abandoning the Tehran and Tripoli agreements by unilaterally raising the posted price by 70 percent. Non-Arab members of OPEC did not join the embargo and cutback strategy, but they raised their prices. Although the production cutbacks were reversed some months later (January 1974) and the embargo subsequently dropped, prices were not only maintained but increased substantially—in some cases reaching $11.50-16 per barrel.[a] Prices were fixed for the first quarter of 1974, and this resulted in the scrapping of the Geneva agreements, since prices were not reduced as the value of the dollar appreciated.

The Recent Course of Oil Prices

Although there have been other elements in rising fuel prices, such as higher materials costs in electricity generation and transmission, supply constraints in the Western low-sulfur coal industry, higher wages, and general inflationary pressures, there is little doubt that the piper calling this particular tune has been the OPEC cartel. The "take" of the oil-producing countries from a barrel of oil in 1970 was less than $1. The agreements between the oil companies and the

---

[a]For example, at the peak of the crisis in January 1974 the prices of Arabian light crude, Libyan 40° crude and Venezuelan Oficina 35° were $11.65, $15.77 and $14.25 per barrel respectively. The delivered price of West Texas 36° sweet crude at Philadelphia was only $6.14 per barrel (Foster Associates, *Energy Prices, 1960-73*, Cambridge, Mass.: Ballinger, 1974). However, the US prices reflected the influence of price controls.

producing countries in the next three years led to an increase in the posted price to $1.99 per barrel. In October 1973, during the Arab-Israeli war, the Arabs and Iran raised the price to $3.44. Prices were then pushed up very rapidly, until the price of a barrel of Arabian light oil reached $10.16 per barrel by late 1974. Prices, in fact, vary quite a lot, partly because of differences in quality and location (access to markets), partly because of deals and contracts, and partly because of the nature of the price system. The posted price that has historically dominated the headlines is, in fact, a hypothetical price which no one pays. More important have been the "equity" price, computed on the basis of the posted price and composed of taxes and royalties, paid by the oil companies to ship out their oil, and the "buy-back" price, a higher percentage of the posted price, which the oil companies pay in order to buy the government's share of production.

In January 1975 a new base price system was introduced in which market prices and government revenues on other grades of oil from other countries were linked to the price of an average barrel of Arabian light crude shipped out of the port of Ras Tanura. At the time of writing this price is $10.46 per barrel. Although there has been some shaving of prices on some deals in view of the surplus of production associated with the stagnation of demand induced by the combination of rising prices and economic recession, this reference price still stands. In fact, oil prices have approximately quintupled since before the war of 1973. The impact on domestic fuel prices has been modified by the retention of controls on natural gas prices and on "old" oil (restricted to $5.25 per barrel), though the FEA has announced its intention to eliminate these controls over the next two or three years. Thus, average electricity bills in 1974 rose by little more than 20 percent, gasoline prices rose by one-third between September 1973 and January 1975 (peaking within this period around mid-1974) and the price of home heating oil rose by almost 60 percent.

The increase in the world price of oil is an index of the short-run monopoly power of OPEC. It bears no relation to production costs, perhaps 12 cents per barrel in Saudi Arabia, 60 cents in Venezuela, and $2.50 or more in the United States. The justifications used by OPEC to explain the rise in prices do not stand up to scrutiny. The oil companies have undoubtedly profited considerably from the recent price rise, and will continue to do so, but most of their profits accrue from their oil operations elsewhere. However, their share in the price of OPEC oil averages less than 50 cents per barrel, out of which their risks, costs, and investments must be recovered. The other OPEC pretext has been the argument that the inflation of world prices of manufactured goods has escalated their import bills. To put this into perspective, since 1970 world export prices have risen about 60 percent while OPEC income per barrel has increased by 1000 percent.

The Impact of the 1973-74 Oil Embargo[1]

The Arab oil embargo of 1973-74 provides some evidence for an assessment of the vulnerability of the United States to interruptions in energy supplies. Unfortunately, any attempt at an evaluation is bound to be defective. Ideally, what should be compared is the actual experience during the boycott with what would have happened in its absence. This cannot be done accurately, because it is difficult to isolate the effects of the embargo itself. A main source of trouble is that none of the forecasting models that were available at the time includes an energy sector specifically, and hence cannot account adequately, if at all, for its interrelationships with the rest of the economy. With the benefit of hindsight it is clear that the forecasts referred to at the time overstated the level of economic activity that would have prevailed in normal circumstances, and hence tended to overemphasize the impact of the embargo. The economy was sliding into recession, a tendency not fully reflected in the forecasts. However, if corrections are made for this, one problem is that the longer-term impacts of the embargo may have helped to induce the recession, particularly the nonquantifiable psychological effects on business expectations. Another difficulty is that the embargo was accompanied by an unparalleled increase in crude oil prices, and the embargo effect has to be separated out from the price effects. Moreover, the embargo brought home the evidence of an underlying change in long-term trends in the energy sector, namely the transition from a buyer's to a seller's market. (The accuracy of this description depends on the price level. If sellers set the price too high, excess supply symptoms may appear.) Accordingly, behavior in energy markets in the winter of 1973-74 reflected changing responses to the longer-run situation as well as to the embargo itself.

In spite of these analytical difficulties, it is possible to make some general statements about the impact of the embargo. The shortfall, as measured by the gap between unconstrained demand and consumption (domestic production, imports, and decline of inventories), widened from 1.1 MBD in the fall of 1973 to 3.5 MBD by February 1974. By the latter date, imports of many petroleum products—especially crude and fuel oils—were down by more than 20 percent from February 1973. The aggregate effects on the economy were quite marked: output in the first quarter of 1974 was $10-20 billion lower than might have been expected in the absence of the embargo; the unemployment rate was raised by about 0.5 percent, or over a half million jobs—perhaps up to 40 percent of the total in directly affected sectors, such as gas stations and airlines, the remainder in indirectly affected industries, such as automobile production and hotels; the contribution of higher energy prices to inflation is reflected in that they accounted for at least 30 percent of the rise in the consumer price index.

To consider the industry effects in a little more detail, the automobile industries suffered most. Four-fifths of the industrial layoffs were related to the decline in the demand for cars and recreational vehicles. This implied that the Midwest accounted for two-thirds of energy-related unemployment, more than 70 percent of this in Michigan. Small cars sold much better than larger cars, unlike the situation in 1974-75. The energy-intensive industries showed few signs of suffering, though the embargo brought home the need for conservation. The service industries related to cars were depressed: 26,000 car dealerships failed and 60,000 (about 25 percent) gas stations either closed down or changed hands. The motel-hotel industry experienced a decline in occupancy rates—in some cases up to 65 percent—while total room revenues fell by about $180 million.

Consumption was affected by the allocation program, by higher prices, and by uncertainty about future prices. For example, average household temperatures declined by about 2° Fahrenheit. The impact on car travel was uneven. Work trips by car hardly fell at all—there was not much shift to car-pooling, and public-transit ridership gained very little. On the other hand, social and recreational trips fell substantially, probably more because of uncertainty of supply than higher prices. The reduced speed limits had some impact on fuel consumption, but a more significant benefit was the decline in the accident rate (an estimated saving of about 4800 lives between November 1973 and April 1974). Finally, lower consumption meant a loss in government petroleum tax revenues of $700 million, substantial but smaller than anticipated in some quarters.

In retrospect, apart from the inconvenience of gas station queues, the adverse effects of the embargo on the economy, though far from negligible, were tolerable. One factor may have been the government's allocation and conservation program. The strategy aimed at minimizing adverse growth and employment effects by accommodating industrial needs, thereby forcing the sacrifices on to the private consumption sector. This was achieved by encouraging voluntary conservation, by reduced availability of retail gasoline and home heating fuels, and by a regional allocation program. All petroleum products were allocated under the Emergency Petroleum Allocation Act of November 1973 from refinery to end-user, apart from gasoline, which was allocated only to the wholesale level. A gasoline-rationing scheme had been prepared by January 1974, but it became unnecessary to introduce it. The major aspects of the regulations that were employed were: priority to food, defense, emergency services, and fuel production; designation of 1972 as the base year for consumption and allocation; and limitation of the states' role. However, the absence of strains was in large part due to the relatively short duration of the embargo and the relatively small size of the shortfall.

There may be longer-run effects of the embargo, but these are difficult to identify in the absence of relevant forecasting models. The forecast errors of the standard quarterly models increase over time. Moreover, in the longer run high energy prices will have influences on final demand mix and on the structure of production. No model is capable of dealing with this adequately, though the Hudson-Jorgenson model does make input-output coefficients a function of prices (see p. 6). One result of the experience of 1973-74 is an appreciation of the need for a better forecasting model with an explicit energy sector and supply as well as demand components, and capable of generating long-term impacts. The other, more important implication of the embargo experience is the need for and the benefit of developing strategies to deal with the risk of recurrence of an interruption of supplies. Elements in such a strategy, emergency stockpiles for example, have to be planned in advance rather than as an immediate response to a developing situation.

Understanding OPEC

The activities of OPEC since late 1973 have created a diversity of hysterical reactions on all sides, most of them based on spurious arguments. Some complaints of the developed oil-consuming countries are that high oil prices have been the major source in accelerating inflation; transfers of funds from the consuming to the producing countries have distorted the operations of the international monetary system; and the behavior of OPEC runs the risk of plunging the world into a severe, possibly inescapable depression. The impact on inflation is undeniable, especially in view of the short period over which the prices were raised. But the monocausal explanation does not stand up to scrutiny. World prices had already started to accelerate beyond the trend rate of inflation before the oil embargo and the price increases, especially for food and raw materials. Between 1973 and early 1975 food prices were as powerful a propelling force as oil prices. Moreover, although rising energy costs affect all sectors of the economy, the strength of the impact differs widely from one sector to another. The implications for transportation and utilities, for example, are much greater than for sectors that are not direct users of fuel. Furthermore, inflationary forces have reached the stage at which rising prices for components in the commodity mix have become less important than the role of expectations among price setters who increase their prices much more than justified by objective costs to protect themselves against *expected* future inflation. This self-feeding characteristic of the inflationary spiral is much more difficult to cope with than a "one-shot" injection, however great its impact.

The disturbance to the international monetary system thesis is even weaker. True, the distribution of reserves changed markedly as a result of rising oil prices. For instance, IMF data show that the thirteen OPEC countries held 19 percent of the world's gold and foreign exchange reserves ($38 billion) at the end of September 1974, compared with 7 percent ($13 billion) a year earlier. The shares of countries such as Great Britain, France, Italy, Japan, and the United States declined. Over 1974 as a whole OPEC built up a surplus of almost

$60 billion, more than one-half of their total earnings. These shifts aggravated individual balance-of-payments problems, partly offset in some cases by capital inflows from the oil-producing countries, but this hardly implies a breakdown in the international monetary system. On the contrary, because of conservatism and inexperience, the oil producers tend to use their reserves in a way that fosters stability rather than instability, in the sense that their deposit shifts tend to be less volatile than those of their more return-conscious predecessors. The range of banks used for deposits is much wider, including some in Japan as well as the traditional outlets in the United States and Europe. Their predilection for short-term deposits is more a reflection of conservatism and suspicion of the host countries than an intention to make quick, large-scale withdrawals. Moreover, some of the oil producers—notably Saudi Arabia and Iran—have lent money to western governments. In former times, monetary transfers on this scale might have been expected to create a serious "transfer problem." There are few signs that this has been a difficulty in this case. The reasons include the multilateral settlements network, including the activities of the IMF; the willingness of oil producers to deposit oil receipts in consuming countries; and the willingness to engage in bilateral trade with some of the main consuming countries—for example, oil for arms in US-Iranian trade. The capacity of OPEC countries to absorb imports from the rest of the world, but especially from the industrial countries, has turned out to be greater than anticipated.

The OPEC-induced world collapse hypothesis was a creature of the media and the politicians rather than a serious matter for analysis. It was, in part, an extension of the arguments about inflation and the collapse of the world monetary system, but it was nourished by the coincidence of OPEC action with a world recession, severe economic and social readjustment problems in many domestic economies in developed countries, reinforcement by OPEC propaganda about western decadence, and a few crisis-ridden countries such as Italy. Rising energy costs induced considerable uncertainty in the high-consuming countries and forced new kinds of readjustment. The result was a crisis of confidence which was reflected in rates of inflation, stock market activity and, to a lesser extent, in business expectations. These primarily psychological factors had a stronger impact because of their synchronization with economic recession. Nevertheless, examination of economic indicators—GNP, investment, unemployment—suggests recession, *not* depression. The post-1929 era could have been recreated only by acts of stupidity by amnesic governments. It is too easy to underestimate the adaptability of capitalism. In the unlikely event of world economic collapse, its roots would be very complex—much too involved for OPEC to be considered its main agent rather than a minor symptom. "Cut oil prices and save the world" was not likely to persuade the Shah of Iran, but its value—if any—lay in its propaganda appeal, not as a serious policy prescription.

A milder, more detached, and more optimistic interpretation of OPEC activities favored by some economists is to treat them as the operations of an economic cartel, directly analogous to an oligopoly cartel in an individual industry. This view uses standard economic analysis to explain OPEC's past and predicted behavior. It points to the tremendously wide gap between world oil prices and Persian Gulf production costs, and—less confidently and less convincingly—to an excess of *potential* rates of extraction over prevailing demand levels. The argument is that cartels are unstable in this situation because one or more sellers could raise profits by breaking away from the cartel, selling more at a somewhat lower price, and generating a price-cutting war. However, the parallel with a typical cartel is misleading. As the experience during the embargo shows, many of the members of OPEC have been motivated by political reasons, and money profit may not have been—at least consistently—the dominating consideration. Cartels are more vulnerable to disintegration when one or more efficient members are being squeezed or unduly restricted by the operations of the cartel, particularly when production levels are held down well below suboptimal volumes. This is a much more serious problem in manufacturing industries with heavy fixed costs than in the oil extraction industry, where oil in the ground can be expected to make capital gains, so that output restrictions are not burdensome. Moreover, the profits accruing to all members of OPEC are so large that there is little incentive for any member to risk disintegration of the cartel by taking unilateral action. As long as the consuming countries are willing to pay current prices there is nothing to be gained and much to be lost by withdrawing from OPEC. A break-up is within the realm of the possible, but would be much more likely to result from political than from economic motives (disagreement between Arab and non-Arab members about goals, or revival of conflict between Iran and Iraq). Industrial cartels break up because of internal or external economic pressures. In current or immediately foreseeable circumstances, and despite recent price shavings by some members, there is no evidence that OPEC is subject to major pressures.

OPEC attempted to justify its price increases and output restrictions in terms other than charging what the market will bear and political motives. Favored justifications included repayment for years of exploitation by the multinational oil companies; action to keep pace with the inflation of the prices of manufactured exports from the developed countries; part of the redress of developing countries against many decades of colonialist exploitation by the developed countries. All these arguments are weak. Prior to the last few years it is arguable that the multinationals operated oil concessions in OPEC countries very much in their own interest, and that the governments of these countries failed to extract the maximum benefit from their ownership of a valuable natural resource. However, as examination of oil company balance sheets for 1973-74 shows, the oil companies have gained more out of the new OPEC strategy than they ever did in their years of "exploitation." The OPEC countries and the oil companies have been co-beneficiaries from a policy that squeezed the oil-consuming countries. The inflation argument is also suspect. Oil prices have increased many times faster than the price increases of world manufactured goods. The very favorable balance-of-payments positions of the oil producers show that their imports of world manufactures have been far smaller than their oil exports. An inflation-minimizing objective, both in respect to world or domestic prices, would have suggested a quite different price strategy for OPEC oil. That OPEC represents the developing countries in their stand against the developed countries is the least convincing pretext of all. Only a small minority of developing countries are oil producers. In general, they have very poor resource bases, and despite their low energy consumption per capita levels they rely more (in terms of import-consumption ratios) on oil imports than most developed countries. Their limited export capacity and capital shortage means that they are poorly placed to pay for higher world oil prices. Also, their ability to substitute other fuels, both in the short and the long run, is generally much lower than in the developed countries. Furthermore, they suffer indirectly to the extent that balance-of-payments problems in the developed countries lead to restriction of nonoil imports and to cutbacks in foreign aid.

Neither the justifications offered by OPEC for its policies nor the objections raised by the developed countries have much merit. Nation states invariably act out of self-interest, in a few cases slightly sweetened with a taste of altruism, and the stances adopted on the issue of world oil prices are no exception to this generalization. Nevertheless, it is debatable whether the policies pursued by OPEC are consistent with the long-term interests of the member countries, though the uncertainties of the distant future make a dogmatic conclusion on this point impossible.

There is a small body of literature in economic theory concerned with determining the optimal rate of depletion of an exhaustible natural resource for a national economy.[2] The value of this type of work has limited relevance in this context. The assumptions behind the models used—such as that resource revenues are used internally, an inverse relationship between the rate of return to capital and the capital-labor ratio and constraints on the variability of the savings ratio—vary substantially from the conditions prevailing in OPEC countries. Accordingly, the central finding of such models—that the optimal exploitation rate involves the restriction of output much below the static profit-maximizing level—has to be treated with caution. Moreover, this analysis does not consider the case of a supplier (or cartel of suppliers) accounting for a substantial share of world supplies, and hence ignores the implications of restriction on energy conservation, research and development into alternative fuels and fuel-saving technologies, and interfuel substitution. These observations are not intended as a criticism of this interesting new area of economic theory, but are merely a reminder that the predictions of these models cannot be used in support of current OPEC strategies.

OPEC countries cannot be treated as a homogeneous set—even from the narrow point of view of determination of the optimal rate of extracting and pricing their oil. Leaving aside political influences such as ideology, political stability, and the time horizon of the government, the two major economic factors are the stock of reserves and the individual country's capacity to absorb oil receipts profitably. These conditions vary widely among OPEC countries. These differences in interests do not imply a break-up in the cartel, merely that there is no optimal rate of extraction. If a common price is fixed, with consequential production limitations, this price must either represent a compromise between the optimal prices for individual countries or imply a strategy that yields widely divergent benefits between member countries. If the resource is exhaustible and absorptive capacity for oil receipts limited, the economic case for some output restriction is very strong. The political embargo of late 1973 was so successful because the policy also contained more than a grain of economic sense. The real problem is how to determine the appropriate degree of output restriction or the cartel price which maximizes social benefits to the member countries in the long run. This choice must take account of the possible responses in oil-consuming countries to a high world oil price.

Projected energy demands relative to known and potential oil reserves suggest that all OPEC oil is marketable, and there must be some price and output path that is optimal in terms of maximizing the sum of the streams of benefits over time discounted at the appropriate rate. The main dangers for producing countries are to run out of oil too soon by flooding the market at low prices; conversely, to raise prices and restrict output so much that oil is left unused in the ground when the switch to alternative fuels occurs; to use oil receipts in a wasteful, profligate, or otherwise unproductive manner; and to fail to develop and diversify the economy in a way that permits structural adaptability when oil reserves run out. (This list excludes the risks of military intervention, either directly by oil-consuming countries (especially the United States) or, more plausibly, via surrogates, possibly within OPEC itself (e.g. Iran). This hidden threat, intended to act as a moderating influence on price-fixing, appears more often in the media or in the fantasies of armchair military strategists than in official statements. Apart from the dubious morality, the dangers of political backfire are so great that this possibility may reasonably be discounted.)

Although it is very much a matter of subjective judgment, because so many of the parameters in this complex situation are unknown, it is arguable that current OPEC prices are too high, even from the point of view of the national self-interest of its member countries. The clear success of the OPEC strategy in 1973 was largely due to the inflexibility of the economies and life styles of the consuming nations; elasticities of demand were so low in the short run that they had little choice but to pay up, and the Hobson's choice was reinforced by the absolute scarcities consequent upon the embargo. The maintenance of these prices over a long period, however, could lead to quite different results. Upward shifts in the price level on this scale induce many effects: lower demand (e.g. electricity consumption); a social environment more favorable to energy conservation, whether voluntary or mandatory; the search for new discoveries of oil and other fuels, both at home and abroad; the substitution of other fuels; research into energy-saving technologies; more research and development in alternative fuel technologies such as nuclear, oil shale, and coal and gas liquefaction; and an import-restriction strategy. None of these responses are likely to be successful in the short run. Some of them may have minimal impact even in the long run. But the stimulus of high prices to technological research could easily lead to a major breakthrough adversely affecting the demand for oil to a drastic degree. It is difficult to assign probabilities to events of this kind, but a lower OPEC price would substantially reduce such risks without serious adverse effects on the oil-producing countries.

This last observation is based on the argument that the marginal utility of additional oil receipts to OPEC countries is rapidly declining. Their capacity to absorb oil receipts domestically in the form of capital is limited. This is not because of the lack of domestic economic opportunities, at least in some countries such as Iran, but rather because of crucial bottlenecks in OPEC economies which restrict the effective use of capital and result in strong inflationary tendencies. Such bottlenecks include human resource deficiencies, e.g. the shortage of technical and administrative personnel, scarcities of construction materials, and the delays in introducing complementary investments in interrelated sectors. Moreover, the absence of capital constraints frequently results in waste and conspicuous, but socially dubious, expenditures on prestige projects or armaments. A sensible alternative is investment abroad, though opportunities there are made less profitable as a result of higher fuel prices. Again, there may be an optimal price level to maximize investment returns, but how this price level might be estimated is very difficult. A further possibility is to continue to build up short-term foreign balances in overseas banks. Not only may this eventually depress the rate of return on such balances, but it is unclear whether this is a more profitable outlet than a pricing policy that interferes less with growth prospects in the developed countries and that discriminates less against the resource-deficient developing countries.

Finally, it is doubtful whether large inflows of receipts create conditions favorable to structural change in the economy. Diversification of the oil economy requires the development of export substitutes. If oil receipts were invested wisely, this could make the promotion of new export sectors easier. But plentiful capital may make it less taxing to rely on imports for manufactures, may divert resources to less socially compelling but more attractive sectors, and may impede international competitiveness via its inflationary effects. Moreover, the social constraints on development may be accentuated rather than relieved by conditions of capital abundance. Of course, these qualifications scarcely alter the fact that the oil-producing countries are far better placed than the other developing countries from all points of view. If oil resources bring their troubles, these are the kind of troubles that most countries would love to have.

Strategies

The events of late 1973 emphasize that the world oil market is not understandable in terms of economics alone, but involves international relations, high-level diplomacy, political interests, and international finance. The implications of the embargo and the price increases for the long-term future of Middle Eastern oil are open to two divergent interpretations. One view argues that the huge gap between the current posted price and the long-run marginal cost of producing oil in the Middle East must induce rapid expansion in supply, generating conditions in which OPEC unity will break up and prices eventually fall. This scenario requires two important preconditions: the replacement of the multinational oil companies—the unwitting commissars of the cartel—by possibly less disciplined national producing companies; and toughness on the part of the consuming countries in their dealings with OPEC.

The opposing view argues that demand is growing rapidly, and that the appropriate supply curve is not the long-run curve but a short-run curve shifting over time. Prices can remain above long-run marginal cost indefinitely, because producers can deal with temporary excess capacity by waiting for demand to catch up rather than by competitive price-cutting. Also, in conditions of increasing demand restricting production is good for revenue. Economic gains are matched by national political interests, and—so the argument runs—these forces are strong enough to hold OPEC together.

The MIT economist, M.A. Adelman, is the strongest supporter of the first view. He sees the events of 1973 as a tragedy of errors in which the oil companies and the consuming countries (and especially the United States government) played into the hands of OPEC by helping their strategy to work.[3] The consuming countries built up the confidence of OPEC by exaggerating the crisis, asked for supplies at almost any price, and "surrendered without a fight." Defensive policies such as rationing were either not adopted at all or imposed too late, the consuming countries refused to cooperate in pooling but preferred to try to arrange bilateral deals with Arab producers behind each other's backs, and they allowed the embargo to become a great political success by giving way very quickly to the Arabs' political demands. The economic success of the embargo was much more debatable, since the decline in US imports amounted in December 1973 only to 2.4 percent of total supply. Nevertheless, the cartel's activities had substantial harmful effects: much higher import bills and consequent balance of payments difficulties; disturbance to the international monetary system; disincentives to mineral exploitation by multinationals in developing countries because of the higher political risks; conflict about ocean reserves and their ownership; and the build-up in arms in the Persian Gulf. A compensating benefit is export expansion and development contracts for the industrialized countries.

OPEC's "success" in 1973 will make it difficult for the consuming countries to devise an effective policy for dealing with the cartel, but the following strategy should be tried. First, the oil companies should cease to be the price-fixing agent of OPEC, and nationalization should be welcomed as being in the interests of the consuming countries. Second, imported oil should be controlled and taxed. Third, the consuming countries should "do and say nothing" rather than organize themselves and make a new Tehran-like deal with OPEC. Fourth, steps should be taken to insulate the US economy—the key strategic point—from Arab oil sources, either by barring imports from countries declaring embargoes or by making contracts with countries interested in preferential entry.

Looking beyond the clear anti-OPEC bias of Adelman's analysis, its basic premise is that long-run marginal costs must ultimately exert their influence on price and output. If this premise turns out to be incorrect, his prescriptions will not work, and an alternative harder road must be found. One possibility is a storage program (see pp. 132-36). This gives a reasonable degree of protection against the effects of an embargo at modest cost, though it has the defect that the producers will probably know how long the store lasts. Another strategy—dangerous in Adelman's world—is coordination among the consuming countries. Walter Levy, for instance, argued for a consuming bloc (European, American, and Japanese) to determine joint policies on bargaining with producers, development of alternative sources, rationing and defensive measures for times of scarcity, and stockpiling schemes. To fight the oligopoly with an oligopsony is appealing, but there are difficulties. There might be a higher probability of bargaining stalemates. More seriously, in hard times there will be temptations for individual countries to break away. In any event, some countries with reasonable prospects for future supplies—such as the North Sea countries—may be reluctant to participate, in case the agreement results in pressure to share out their own domestic supplies.

For the pessimists the best hope—perhaps the only hope—is to restrain the growth of demand. The difference between a 4 percent and a 3 percent demand growth might make an impression, especially when Alaskan, North Sea, and other non-OPEC fields start to flow. But even 3 percent growth will not have much of a downward impact on price. Also, even allowing for new discoveries on a substantial scale, two-thirds of world oil exports would probably come from the Middle East and over one-half from Saudi Arabia, so that much depends on the output strategy of the Saudis.[4] If demand growth could be restricted by strong energy conservation measures, including high gasoline taxes, if non-OPEC petroleum resources can be expanded and brought on to the market rapidly, and if the research into alternative fuel sources and new fuel technologies pays off quickly, it may be possible to erode the power of OPEC. But these are big ifs, and the pessimists believe that the dynamics of demand and supply will favor the suppliers and the continuation of the cartel for a long time to come.

Although one hopeful sign is the slowing down of growth in demand for

OPEC oil, reinforced by higher prices and explicit policies in the major consuming countries to reduce dependence, it would be wrong to expect the situation to revert to pre-1973 days. The OPEC countries, including—perhaps especially—the non-Arabs, have clearly assumed total control over the official price level and their choice of customers, as well as developing a degree of self-discipline in restricting production. The United States' response to changing conditions was a new strategy: an attempt to secure a continuous flow of oil at lower prices than those prevailing since late 1973; a longer-term policy of undermining OPEC supply control by developing alternative fuel resources; and cooperation among the oil consumers to counteract the cartel, to coordinate energy policies, and to prepare defensive measures against a revival of the embargo.

This strategy has far less appeal in Europe. European countries have less to fear from an embargo related to the Middle Eastern political situation, because they can more easily accommodate the Arabs, and lack the commitment to Israel. However, they have more to lose from an embargo motivated by an economic struggle between producers and consumers. At present, they are much more dependent on Middle Eastern oil than the United States, and have fewer domestic resources to fall back on in periods of shortage. In addition, some countries such as France are unwilling to accept a tutelage relationship with the United States for promotion of a strategy largely in the interests of the latter. The alternative policy of bilateral deals guaranteeing flows of oil at the prevailing market price but connected with large capital goods export contracts had considerable appeal. However, attempts by France to seek a "special relationship" have been unsuccessful.[5] Such deals are likely to be based on narrow economic interest, and to remain restricted to that sphere. The barter of oil for goods does not guarantee the flow of oil, nor does it ensure that the bargaining power on both sides will be equal. Also, it offers little prospect of prising apart the OPEC cartel, since the members realize that they could not have achieved alone what they achieved together. Of course, there is a problem of policing output restrictions to keep prices high, but this should be manageable. Moreover, the threat of the consuming countries to make themselves self-sufficient is much emptier than appears at first sight, since by the 1980s the producers will have built up enough revenues to live off interest alone, and hence may be relatively indifferent about the market for oil.

Any assessment of the scope for an agreed international energy policy among the major industrial countries must take into account the more vulnerable position of the European countries and of Japan compared to the United States. These countries generally lack domestic oil, gas, and coal reserves on the scale of the United States, and the costs of government-financed nuclear power will impose heavier strains on their budgets and resources. Although the United States is easily the largest oil consumer—consuming more oil than all of Western Europe put together—and although imports were growing rapidly in the years prior to 1973, nevertheless her dependence on oil is lower than that of the other major industrial countries, her petroleum consumption has been growing at a slower rate, and her relative dependence on Middle Eastern sources for oil imports is also lower. Japan, for instance, relies on oil for over 70 percent of her primary energy resources, petroleum consumption rose at an annual rate of 17 percent in the decade 1962-72, and three-quarters of her imports come from the Middle East. Western Europe is 60 percent dependent on oil, and two-thirds of its imports come from the Middle East. European oil consumption in the 1960s increased at double the rate of that of the United States. The limited freedom of negotiation of Western Europe and Japan and their lack of bargaining power in the short run must be taken into account in explaining their hesitancy and lack of enthusiasm for a joint stance against OPEC as urged by the United States.

The longer-term prospects are somewhat brighter. Energy conservation strategies may pay off. Exploration for new oil reserves is already yielding results. The specter of immediate physical shortage will recede and the economic strength of the concentration of most reserves in the hands of a small group of countries will diminish. By 1980 the viability of the OPEC cartel may be suspect, and indeed its rationale may have disappeared. The eventual weakening of world oil prices is important, not so much for the survival of the industrial world—which by that time will have adapted—but for those parts of the developing world with limited access to domestic energy resources. However, this scenario depends not only on the rapid exploration and development of new oil and gas fields but on the assurance that these new supplies reach the market. The oil companies may find it profitable to go slow on the rate of exploitation in order to share in the high profits associated with high prices. The case for government vigilance and control to ensure that this does not happen is as strong as ever. If the governments of oil-consuming countries baulk at the intervention associated with state-administered oil supply, they should at least be sure that they act as good watchdogs.

Emergency Stockpiles

Apart from relentlessly rising energy prices that have long-term damaging effects, the most serious threat to the economy is an interruption of supplies of petroleum as experienced during the 1973-74 embargo. One way of meeting this threat is to have standby emergency supplies available. Reducing domestic oil production by retaining reserve capacity is one method, but it would be more costly and difficult to administer, and might require even greater imports than a storage program. A preferable course is to store petroleum after production, either in the form of crude and residual oil or as refined products. Storing crude is much more flexible if refining capacity is available, and it is also cheaper. It is this option that will be examined here.

Storage is an alternative, or complement, to tariffs and quotas. Its advantage is that it may offer a similar degree of protection and security against supply interruption at a lower cost. For example, one hypothetical calculation suggested that a storage-tariff option would reduce the welfare loss compared with a tariff alone by 75 percent.[6]

The United States, almost alone among major industrial countries, does not stockpile crude beyond normal commercial inventory needs. This probably reflects the size of the domestic petroleum industry which, prior to 1973-74, appeared to make emergency stocks unnecessary, and the more limited involvement of the government in the energy sector. Normal inventory levels of crude and refined products are within the range of 800-1000 million barrels, or about 15 percent of total annual demand (i.e. about sixty days' supply). Since little more than one-half of this is needed as working stocks, at first sight there would appear to be a sizable surplus to deal with short-term interruptions to supply. However, the product mix and the geographical distribution of stocks make it difficult to use them in an emergency. For example, in 1972 residual fuels stocks equalled 23 days of consumption, crude 24 days, gasoline 34 days, distillates 51 days. Seasonal variations were also wide: gasoline stocks fell from 40 to 28 days between winter and summer 1972, while distillates stocks rose from 34 to 77 days. Moreover, there was a reluctance during the embargo to draw upon stocks, in view of the uncertainties about the length of the embargo.

There are several problems in developing a satisfactory storage program. Some of these relate to the construction facilities. The stores must be located at sites where petroleum can easily be transferred to distribution points or to consumption centers. There are substantial lead-times for the construction of storage facilities or for conversion of natural stores. The costs of different methods vary widely, and this is an important consideration in evaluation of the net benefits of a storage program relative to the degree of protection a stockpile of a particular size offers.

The methods of storage include steel tanks above ground, underground salt domes (either onshore or offshore), abandoned mines, and mined caverns. The last of these is the most expensive ($5-$12 per barrel), and limited to hard rock formations. Abandoned mines are unsuitable, and the relatively few convertible mines are distant from refineries and distribution centers. Steel tanks have the greatest flexibility locationally since they can be constructed anywhere, subject to planning constraints where several-acre sites are available, but they are quite costly ($3-$5 per barrel), and there could be constructional delays due to capacity constraints in the steel industry. This leaves salt domes as the most suitable all-round facility. Costs are modest (construction costs of $0.50-$0.75 per barrel onshore and $0.75-$1.00 offshore, while maintenance costs are negligible); the more suitable sites are near the Gulf Coast so that transportation facilities would be required to deliver supplies to the East Coast during an emergency (hence the importance of the deepwater terminals planned for the

Gulf Coast); the lead-time for several hundred million barrels is about six years, so that a storage program could be in operation by the early 1980s. A single 200-million-barrel project might require twenty to forty wells only, and could redeliver several MBD.

A critical question in evaluating the usefulness of a stockpile strategy relative to other methods of protection is the required size, and hence the cost. Total costs per barrel can be broken up into costs of acquiring the oil, costs of the storage facilities, and maintenance costs. The per barrel cost is then multiplied by the size of the stockpile, which is related to the estimate of insecure imports (in terms of MBD) multiplied by the estimated length of an embargo, supply interruption, or degree of time protection required. Since the volume of insecure imports is affected by the import policy in operation, the required stockpile size will vary widely according to whether imports are unconstrained or whether a protectionist tariff is in operation. It may also be feasible to trade off between stockpile size and acceptable reductions in GNP due to supply disruption. Given the many variables and their unpredictability, it is hardly surprising that estimates of the costs of a storage program vary considerably.

The Brookings Institution[7] estimated that it would cost $12 billion (or $1.2 billion dollars a year) to create a stockpile that by 1985 could replace imports of 6 MBD for one year (the 90-day equivalent would be $3 billion). But this estimate assumes a world price of $6 per barrel, much too low, and takes inadequate account of storage costs. On the other hand, it is doubtful whether protection against imports of 6 MBD would be necessary, if high prices repress import demand and if a proportion comes from relatively reliable sources.

The MIT study[8] estimated costs for two alternative programs, a "low" stockpile and a "high" stockpile. The former assumes dependence of 15-20 percent on imports and the existence of other policies to promote energy independence. In particular, it assumes tariffs and quotas and stimuli to domestic production so as to reduce imports to 4 MBD, one-half from insecure sources. Assuming that it might take one year to replace insecure supplies, this would imply a stockpile of 730 million barrels. Storage costs are assumed to be $3 per barrel (a mix of underground salt domes and above-ground steel tanks), annual operating costs 10 cents a barrel, and the oil itself $8 per barrel. Interest and depreciation are assumed to be 15 percent, while the interest on the oil is assumed to be 10 percent. On these assumptions the total carrying cost of the stockpile would be $1.35 per barrel, implying a total annual cost of $990 million. The construction of storage facilities might cost $2.2 billion, but this could be phased over several years. The "high" stockpile program assumes unconstrained imports, no other energy policies, and a much longer time needed to find alternative supplies. For instance, insecure imports of 4 MBD and a three-year interruption would need a cost of $5.9 billion to maintain a protective stockpile. This could strain the construction industry, and in any event implies an additional cost of 3.6 cents per gallon.

The estimates of the Federal Energy Administration (PIB)[9] are more complex, because they embrace a wider range of possibilities, including an International Energy Program (IEP), combining a stockpile with emergency demand restraints, the impact of stockpile purchases on world oil prices, and other considerations.

PIB assumes acquisition costs of $10.50 per barrel, storage facility costs of $0.85, and maintenance costs of $0.01 per barrel. Assuming annual costs to be 15 percent of capital outlay (on oil and facilities), total annual storage costs are $1.73 per barrel. This sum can be multiplied by desired stockpile size to yield total annual cost, while an appropriate time horizon might be ten years. A second cost element is the GNP lost because of supply disruption; this can be avoided if the stockpile is large enough. A PIB example considers a one-year supply interruption of 4.8 MBD in the context of a total import dependence of 9 MBD. A stockpile level of 500 million barrels (costing $8.7 billion) would involve a decline of $80 billion in GNP; a 1500-million-barrel stockpile (costing $25.9 billion), on the other hand, would avoid any decline in GNP. Total costs would be minimized with a large stockpile, in this case of around 1400 million barrels. If the insecure share is 60 percent, the cost-minimizing stockpile increases from 73-365 million barrels to 2700 million barrels, implying 23-125 to 180 days of imports, as total imports rise from 3 to 15 MBD. Of course, the stockpile costs can be spread over all imports; thus, if a one-billion stockpile costs $1.73 billion per annum, and annual imports are 5 MBD, this works out at a marginal cost of $0.70 per barrel.

Building up a large stockpile creates problems in the form of greater risks of capital loss on the cost of oil purchase and the impact of additional demand on world oil prices. For example, PIB estimated that acquisition of a small stockpile of 144 million barrels (3.2 MBD for 45 days) could raise the cost of all US oil imports by $542 million (implying an additional 5 cents per barrel cost of the oil in storage). If Western Europe and Japan were simultaneously building up storage, the additional storage cost might be as high as 20 cents per barrel.

Accordingly, there are incentives to minimize the size of the stockpile as far as is compatible with security. If there was a shortfall of world oil exports and the United States drew upon its stockpile to the same extent, thereby eliminating the world shortage, no increase in world oil prices would result. However, all the burden of supply interruption would have fallen on the United States for the benefit of the rest of the world. The costs could be reduced considerably via an IEP (International Energy Program) including diversion of imports to areas of need, sharing of stockpile reserves, and possibly even the sharing of domestic production. Also, an IEP might contain provisions for joint demand restraint, for example up to 10 percent of consumption, in periods of supply disruption. A ninety-day IEP import stockpile supplemented by a US domestic stockpile might be a very cost-effective strategy for protection against unforeseen embargoes, even if a severe embargo occurred only once every decade. Another measure could be mandatory emergency demand restrictions; PIB estimated that cutting down demand by one MBD was feasible. Assuming a one-MBD conservation strategy, the costs of a 10 percent demand restraint for three months at an oil price of $9 would be about $5 billion in 1980. The costs of a supplemental stockpile, again at a 1980 price of $9 per barrel, and assuming that the stockpile equaled one year's supply at a daily rate of 10 percent of consumption minus the one-MBD-stand-by conservation, would be $3.8 billion.

As these varied estimates suggest, the costs of emergency stockpiles are far from negligible, and their development creates problems. On the other hand, their relative attractiveness is considerable when compared with alternative methods of achieving the same degree of protection against short-term interruptions in supplies. For example, the MIT calculations at a world oil price (delivered US) of $9 show a high-stockpile price of $10.56, compared with an import-prohibition price of $13 per barrel. The difference of $2.44 means a saving to consumers of 6 cents per gallon. Also, US domestic reserves are conserved with a stockpile, compared with the import-prohibition case. In other developed countries the government requires the oil companies to maintain stockpiles. If this practice were adopted in the United States, the costs would fall upon the price of gasoline rather than upon the public exchequer. The MIT "low" stockpile would involve an increase in the price of gasoline of only 0.67 cents per gallon. The administration of release of supplies in a private-sector program should not be too difficult. Individual industries with particularly large or volatile demands might be encouraged to maintain their own stockpiles. If there were futures markets in petroleum, speculators could perform the storage function, but at present there are no such markets, and little likelihood that they might be developed.

9 Demand Restraint

Introduction

Of all the alternative strategies to deal with the energy problem—such as stimuli to increasing supply, research and development into new technologies, and international cooperation to pool and share supplies—none has a greater payoff, at least for the United States, than economies in energy consumption. There are several justifications for this view. First, the cumulative multiplicative property of exponential growth ensures that it makes a great deal of difference to energy consumption in, say, the year 2000, whether demand grows at a zero rate or at a rate of 2, 4, or 8 percent. Second, if the difficulties of 1973 and after are largely due to restrictions of supply by Arab oil producers, the one sure, effective way of dealing with this, regardless of how well OPEC holds together, is a countervailing restriction of demand. Third, the differential in per capita energy consumption between the United States and the rest of the world is so great (a ratio of approximately 8:1) that the scope for economies must exist, certainly in the long run, and—perhaps more important—economizing by Americans could be interpreted as a concern for international equity considerations. Fourth, it is *not* argued that energy savings confer some moral uplift, or that a Spartan way of life is inherently good. Demand restraint in energy is compatible with narrow views of self-interest: it saves resources and productive factors for other uses; it reduces the disruption resulting from embargoes and other interruptions to supplies; it buys time for research and development into technologies using less scarce fuels; and it reduces negative environmental impacts. Moreover, in the longer run as energy-intensive automobiles, appliances, homes, buildings, and plant become obsolescent and are replaced, massive economies are feasible without any dramatic change in standards and styles of life.

An important characteristic of demand restraints is that, with the possible exception of savage increases in price, they are effective only in the long run. For example, the changeover from large and heavy automobiles to vehicles of European size would take a decade of replacing a stock of over 100 million cars, even if there was no consumer resistance. The time horizon for achieving an energy-saving housing stock is much longer. Even simple measures, not limited by technological constraints, such as persuading consumers of the importance of energy economies, cannot be effective overnight—the process of education and adjustment takes time. The success of oil restrictions in 1973 was due to the short-run inelasticity of demand. The only feasible actions were voluntary restraint, high prices, and rationing, and these were all palliatives. Higher electricity prices in 1973-74 dampened the growth in electricity demand. How elastic the demand for energy is in the longer run is unknown, because there is too little experience of substantial price rises. Also, it is probable that the long-run elasticity is a parameter subject to influence by government action, industry behavior, and the public-relations activities of consumer groups.

Some people argue that the most effective instrument for restraining demand is high prices. This argument is frequently extended to the view that the prevailing high prices are to be welcomed as an encouragement to the reduction of waste. This position implicitly involves assumptions about demand elasticity. It also ignores the adverse effects of high, and rising, prices on consumers, the contribution to inflation, and the equity impacts on the poor.[a] Because of the absence of a substitute for oil in transportation and the huge share of transportation in total energy demand, oil becomes the price-setting fuel. Thus, all fuels tend to adjust to some degree and with various time lags to the world price of oil. If the high-price thesis is valid, OPEC is seen as an unconscious benefactor by waking up American industry and consumers to incipient changes in the world energy supply-demand situations. A counterargument is that the price of oil is merely a cartel-determined price, and that the world energy market is still characterized by excess supply. If this perspective is the more correct, then high prices for fuel have severe distorting effects on the economy. Nevertheless, there remains a case for energy conservation if present energy consumption patterns involve waste. The primary danger of artificially high prices is that they might induce irreversible investments and research expenditures in costly and exotic fuels and energy technologies that are not economically justifiable in the long run.

Residential and Commercial

Energy conservation, like charity, begins at home. A study sponsored by HUD in the Baltimore-Washington area[1] showed that major energy savings could be obtained, particularly in new houses but also to a lesser extent in the current housing stock. The use of storm doors and windows would lead to 12 percent savings in primary energy; a 12 percent increase in wall-insulation volume would save 21 percent; a 25 percent reduction in window area would save 32 percent. Other modifications, such as wooden window frames, the relocation of windows, and the installation of vestibule doors, would achieve minor savings. Structural

---

[a]Families earning below $10,000 a year account for 17.5 percent of family income but make 27.3 percent of direct electricity and gas purchases and 22.5 percent of indirect energy purchases. The corresponding shares for rich families (above $25,000) are 33.7, 24.4 and 24.5 percent. The poor family spends only about one-fifth of what the rich family spends on energy, but fuel accounts for a much higher proportion of its income.

modifications to conserve 500 therms—or 19 percent of energy consumption—were all readily available at a cost of around $550. The use of gas-fired rather than electric appliances might achieve savings of 21 percent of primary energy. High-efficiency lighting and improved thermal design of appliances would require technical advances, but these were feasible, and would result in significant economies. Another possibility was technical improvements to recover 50 percent of the waste heat in the furnace flue, and closure of the furnace in off-periods. This might save energy equal to 7.5 percent of consumption, at a cost of less than $200. High-performance air-conditioning using the open-air cycle might cost a little more but could save 9 percent of consumption. Thus, "the annual energy consumption of a good quality, single family residence could be reduced up to 40 percent without affecting the life style of the occupants."[2] These modifications all fell within the scope of current technologies, though minor design improvements and code revisions might be necessary. The estimated cost was $965—a small sum relative to the house price, and one that would be fully repaid by fuel savings within a few years.

Because home heating by electricity is 165-170 percent more wasteful than coal, gas, or oil, some analysts have argued that the use of electricity for heating homes and offices should be banned. Certainly, forecasts of future electricity consumption identify the residential market as the most rapidly expanding sector, with commerce not far behind. Apart from the choice of heating fuel, other diseconomies include the use of incandescent rather than fluorescent lighting, too high levels of illumination and single switches per floor in offices, the inefficiency of most window-unit air conditioners, the high energy consumption of mobile homes (now one out of every four additions to the housing stock); these are just a few of the more obvious examples. The proliferation of electrical appliances stimulated by advertising and promotional campaigns is another major source of demand, not all of which can be justified as essential. (Some information on the number and saturation rate of major domestic appliances is shown in Table 9-1.) Almost every family owns certain basic appliances—refrigerators, radios and TVs, irons, washing machines, toasters, coffee makers. Although many others are much below potential saturation levels, they tend—with a few exceptions such as air-conditioners, freezers, and clothes dryers—to be lower power consumers. The most striking impression of the data given in Table 9-1 is that a limited number of appliances account for most of the power consumption. Appliances having an average consumption of 1 to 5 kilowatt hours per day are—in rising order—dishwashers, television sets, clothes dryers, freezers, air-conditioners, and refrigerators. Although some of these may not be "essential" from a minimum-standards point of view, none could be described as "frivolous."

There is *scope* for substantial energy economies, both in the home and in the office. The problem is how they can be achieved in view of the time adjustments, the education, and the industrial reorganization needed. For

**Table 9-1
Household Appliances**

Appliance	No. (million)	Percentage of Households	Average annual consumption per appliance (kwh)
Refrigerators	62.6	94.8	1829 (14 ft.³, frostless)
Radios	62.5	99.7	50
Irons	62.4	99.5	144
Televisions (b & w)	61.9	95.0[a]	360
Televisions (color)	23.7	37.8	540
Washing machines	57.6	91.9	103 (automatic)
Toasters	57.1	91.1	39
Vacuum cleaners	56.9	90.7	35
Coffee makers	54.2	86.4	106
Frying pans	34.6	55.2	186
Blankets	29.8	47.5	120
Canopeners	27.1	43.2	n.a.
Clothes dryers	25.3	40.3	993
Air-conditioners	23.0	36.7	1389 (room)
Freezers	18.6	29.7	1217 (12 ft.³, frostless)
Dishwashers	14.9	23.8	363
Waste disposers	14.4	23.0	30

[a]Some households have two or more black-and-white television sets.
Sources: Library of the United States Congress, *Energy Facts* (November 1973).
U.S. Bureau of the Census, *Current Population Reports*, Series P-65 (Washington, D.C.: U.S. Government Printing Office).

instance, a major impact on energy consumption would result from abandoning suburban life styles with the single-family detached dwelling, the long commuting, and the low densities in favor of high-rise apartments, reliance on public transport, and industrial and residential relocation to minimize journeys to work via central city living. In New York City per capita energy consumption is only about one-half of the national average. Although there are other reasons (since suburbs in New York State also consume less energy than in the US as a whole), including the limited manufacturing industry and the shunning of the region by energy-intensive users because of the 60 percent higher utility rates, the more important factor is the high-density urban environment of New York City. On the other hand, the slower *growth* rate of energy consumption in New York is mainly due to other reasons, since densities continue to decline at a rate not markedly lower than in many other American cities; these reasons include the lack of population growth and the less vigorous expansion in per capita income.[3]

Any reversion to a more concentrated pattern of urban development could not take place immediately, if at all. The infrastructure and housing stock of the city have adjusted over decades to a suburban, decentralized life style, and because much of this urban capital is relatively new a return to the central city on any scale would take even longer. In any event, there is no evidence that the higher energy expenditures associated with suburban living have reached a disincentive level strong enough to induce many to abandon their preferred suburban living. If such a change were to occur, an unwelcome by-product in the current situation might be a deterioration in environmental quality due to the close correspondence between high densities and pollution, noise, and other nuisances. However, the adjustment time needed for relocation to become effective also buys time for implementation of environmental standards and technological solutions to pollution creation. For example, it has been suggested that even Manhattan traffic could meet air quality standards if the original federal vehicle emission requirements were implemented in the late 1970s.

Transportation

Transportation is a key sector in any effective energy conservation plan. It absorbs 25 percent of total US energy consumption, and is the most inflexible demand sector because of its technological dependence on petroleum (and is a major factor in the high US share of 55 percent of world gasoline demand). More than 80 percent of American households own a car, and more than 30 percent own two or more. Taking into account the growth of automobile use for business purposes, the number of cars registered was 50 percent higher in 1970, than in 1950 and is now two-and-a-half times greater than in 1950—around 100 million, or more than 125 million for all motor vehicles. Annual average car mileage is now more than 10,000 miles. Fuel economy has declined over the past twenty-five years from 15 to 13.2 mpg, aggravated by heavier cars and the inclusion of fuel-consuming extras. Fuel consumption by automobiles has approximately trebled, and now accounts for 28 percent of petroleum consumption and 12.5 percent of total energy use. As for commercial air travel, despite recent setbacks to the industry's growth, fuel consumption is about three times the level of a decade ago.

The plane and the automobile are both highly energy-intensive modes. The plane on the average obtains 21-22 passenger-miles per gallon, and the car 32 passenger-miles, compared with estimates of 80 for intercity trains, 100 for commuter trains, 125 for buses and 200 for double-decker suburban trains.[4] A different set of estimates[5] suggests that for intercity travel planes are only 40 percent as efficient as cars and 34 percent as efficient as trains, in terms of passenger-miles; for intracity travel cars use double the energy of buses, 27 times as much as walking, and 40 times as much as cycling; for freight in terms of energy per ton-mile pipelines, where feasible, are the most efficient, while rail, barge, and other forms of water transport are one-and-a-half times more costly; the truck is 8.4 and the plane 93 times more costly. Moreover, the dynamics of the situation are moving in favor of the more energy-intensive modes. In the 1960s, for instance, rail passenger traffic declined by more than 50 percent, car mileage rose by 50 percent, and airline mileage increased threefold.

A detailed study of transport in the New York region[6] showed that these trends and patterns are replicated even within a more concentrated spatial structure where public transport might be considered to have a relative advantage compared to other areas. The energy-saving modes (bus, subway, and rail) use only 5.1 percent of the raw energy inputs, but they supply 13.3 percent of the travel demand. Yet these are the very modes suffering from falling demand: bus travel declined 14 percent in the 1960s and subway and rail travel 6 percent. On the other hand, air travel to and from the New York region more than trebled. The number of motor vehicles in the region has doubled every two decades since 1930, mainly in recent years because of second- and third-car purchases. In the 1960s the number of cars increased by 40 percent, yet the number of noncar households actually increased. The car and the plane could account for more than half of the region's energy consumption by 1985.

There is thus a wide gap between potential and feasible savings. The public transport that might substitute for the automobile does not exist in many cities, and there is little encouragement in present travel behavior to justify massive new investments in such facilities. Even where alternative modes are available, it is doubtful whether all the tactics of public transit fare pricing such as uniform below-cost fares are effective in prising the commuter out of his car, although they have had some impact in generating new demand.

The scope for economies depends upon the ability to halt the growth of air freight, to achieve fuel economies in motoring, and to divert passengers in both intra- and intercity travel to alternative modes. The first is feasible by regulation; the second is worth aiming for, partly by regulation, partly by changes in tastes, and partly by pricing measures; but the third goal is much more difficult to achieve, because of the poor urban public transit services and the closure of much of the national passenger rail system. Thus, the estimate that 20 percent of energy consumption could be saved if all ground transportation was by rail has little practical value. That Americans could save 1.5 billion barrels of oil per year by 1985 by changing over en masse to fuel-efficient automobiles is a more relevant, though still idealistic, suggestion.

Industry

Energy conservation in industry also has considerable scope, though perhaps less than in the home or in transportation. Since 1973 most major companies have achieved fuel savings of several percentage points by more efficient operation. In the longer run technological changes may bring much larger savings. For example, vacuum furnaces consume only about one-quarter of the energy of older designs. Potentially most important of all would be technical advances in power-plant design and operation (e.g. combined-cycle power plants), which might raise the present average thermal efficiency of 32 percent toward 50 or even 60 percent.[b] An EPA study suggested that potential savings in energy by 1990 could be 30 percent; 34 percent in industry, 27 percent in transport, and 22 percent in the residential and commercial sectors (compare the estimates of Table 9-2).[7]

Instruments

Of all the possible strategies for curbing fuel consumption by automobiles, pricing and tax measures are likely to be the most effective, at least in the short run (see also Chapter 11). In the longer term, it may be feasible to design cars and to retool Detroit so as to develop fuel-economy cars on a large scale, and to adopt policies—such as differential taxation according to engine size and body weight—to persuade motorists to buy smaller cars. These changes would involve reducing size and weight—which are much more important in reducing fuel consumption than emission-control equipment, improving the performance of engines and transmissions, and altering body design so as to lower wind

Table 9-2
Potential Energy Savings by Sectors, 1985

Sector	Savings (Q. Btu)	Percent of Demand
Residential and commercial	2.52	10.0
Industrial	1.05	3.6
Transportation	2.85	13.0
Total end user sectors	6.41	8.4
Reduced losses in electricity generation	2.35	6.0
Increased efficiencies at utilities	0.91	2.3
Total	9.67	9.4

Source: Federal Energy Administration, *Project Independence Blueprint* (Washington, D.C.: U.S. Government Printing Office, 1974), p. 175.

[b]This century has seen major efficiency improvements in electric power production. For instance, new coal-fired plants may have a thermal efficiency as high as 40 percent, compared with only 5 percent in 1900. Nevertheless, it is widely believed that the potential for further efficiency gains is very low without major technological advances.

resistance. In view of the slow response of the automobile industry and the resistance of consumers to change, a transformation of this kind may require interference with consumer sovereignty, either in the form of discriminatory tax measures or controls on the numbers and sizes of cars produced. The payoff in fuel savings may justify such intervention since the automobile share in total energy use rises to about 21 percent when indirect energy uses are taken into account.

Since the policy objective in the energy-saving case is not directly to inhibit car ownership, or even to reduce urban congestion (though this might be a desirable side effect), but to save fuel, the most obvious pricing strategy is to raise the price of fuel via a higher gasoline tax. (Other instruments to raise gasoline prices, such as tariffs on oil imports, have similar impacts to a higher gasoline tax. In this discussion the latter is used as a representative rather than necessarily the most appropriate pricing measure.) Such a tax (in the form of a 25 percent rather than an 8 percent value-added tax) was introduced in the United Kingdom in November 1974, and there has been much discussion in the United States of the pros and cons of a higher federal gasoline tax. How effective this would be depends on the elasticity of demand for gasoline, an issue on which views differ widely. A recent study by RAND[8] suggested that a one-cent-per-gallon increase in price would lead to a one percent drop in consumption, and that relationship would hold even with massive increases in price. (Examples include: 15 cents per gallon price increase, 16 percent fall in demand; 30 cents increase, 34 percent fall; and 45 cents increase, 40 percent decline in consumption.) This implies a price elasticity of demand within the range of -0.5 to -0.67. Using a different approach (fitting a curve to a cross-sectional scatter diagram relating price to consumption over many countries), the OECD suggested that the price elasticity of demand was about -1; the fitted curve was a rectangular hyperbola, indicating that expenditure per capita remained the same as higher prices were exactly offset by lower demand. If these orders of magnitude are not too far off the mark, then higher gasoline prices can save fuel, and by far from negligible amounts.[c] However, the costs of such a policy should not be ignored. In particular, its regressive impacts could be massive. Its effects on the two- and three-car owner might be minimal, regardless of whether the consumer decides to restrict his mileage or to pay out additional fuel costs that remain marginal in relation to his total income. However, many central-city blue-collar workers are dependent on their cars to take them to the dispersed workplaces of the suburbs. Car ownership may be the key instrument for taking advantage of opportunities outside the ghetto.[9] The restriction of mobility implied in lower mileages would be a social cost of this policy. Also, the unemployment impacts in the automobile and related industries and in

---

[c]However, only one-quarter of consumer response to higher fuel prices takes the form of driving less with a given automobile; most of the response is via the long-run decision about which automobile to buy (research by H.S. Houthakker and P. Verleger).

service stations should not be underestimated, even though they might be transitional.

The case for a higher gasoline tax would be much stronger, of course, if alternative public transit was available for the less well-off who could not afford to use their cars as much. This condition rarely holds. Many American cities never had an adequate public transport system. In other cities the quality and frequency of service have declined appreciably in the last two decades as ridership levels have fallen off. Federal subsidies for mass transit schemes have failed to generate successful projects. Even the prestige projects such as the BART (Bay Area Rapid Transit) system in San Francisco have run into teething troubles, and appear to have little hope of diverting automobile traffic on the required scale. There have been some instances where the transition to a lower and a uniform bus fare have increased ridership (San Diego, Atlanta, Seattle), but most of the expansion has been in off-peak rather than commuter travel. Simple traffic management reforms, such as solo bus lanes, can be important in improving speed of service. Schemes for new subways, mass rail transit, monorails, and other elevated transit systems have a very long gestation period, pending consensus on their viability, securing the necessary rights-of-way and construction, that may run to twenty years or more. The unhappy experiences of public transit systems in the past hardly provide the background environment against which new projects may be assessed favorably.

There is much prima facie evidence that diversion from the automobile to public transport will require disincentives to the use of cars as well as low fares and good service on the public facility. The closure of streets to traffic, bridge and road tolls, penal parking charges, and physical restrictions on the number of parking places, not only in public lots but also attached to private office blocks; these are among the easiest measures to introduce, without facing the problem of the political acceptability of direct road pricing. Nevertheless, most of these controls are feasible only in certain CBD areas that cover only a tiny fraction of the territory where public and private transport compete. Accordingly, they would be effective only if they persuade motorists to abandon their cars for commuting altogether rather than merely to switch modes near the edge of the CBD. An even more serious objection is that negative controls on the automobile have to be synchronized with the provision of better public transport services, but the responsibility for each function rests with different bodies. The integration of policies pursued by the municipalities, the private transit companies, the public and semipublic commissions, the federal government and state authorities, is a difficult task which augurs badly for the design and implementation of fuel-conserving transport strategies on these lines.

A more promising tax measure may be to impose a steeply variable tax on the horsepower or size of cars. Small cars can be twice as efficient as United States standard models (800 calories as opposed to more than 1600 calories per passenger mile). The preference for larger cars is primarily a hangover from the past. Today, a much higher proportion of travel is intraurban, families are smaller, and the maximum speed limit is lower. A target of 21 miles per gallon as the standard efficiency goal for 1985 would seem reasonable if imposed soon, and would give time for phasing in and for the retooling of the industry to meet the changed pattern of demand. Such a target would permit a stabilization of gasoline demand, even allowing for the projected increase in the number of cars. Unless tastes change spontaneously, either as a consequence of higher fuel prices or because of greater familiarity with, and appreciation of the advantages of, smaller European- and Japanese-styled cars, the tax levels needed to shift demand to the smaller, lighter cars may have to be very high.

Other measures to reduce automobile fuel consumption are probably much less effective. Lower speeds, for instance, are not a solution, since less than a third of motor vehicle travel is on the major rural highways affected by the 55 mph speed limit. Over one-half of all car travel is on urban streets, where very slow speeds and multiple stopping induce fuel loss. The case for the 55 mph speed limit on nonurban roads, therefore, rests more on its safety benefits than on fuel savings. Car-pooling schemes are notoriously ineffective, even when promoted by special priority lanes, free rides over toll bridges and other inducements. Persuading three-car householders to ride in someone else's car involves surmounting a considerable psychological barrier. Car pooling is associated with too much inconvenience where work places and residences are dispersed, and is frequently redundant where origins and destinations are clustered, because these routes are more likely to be served by public transit. Car-parking restrictions, lot place controls, traffic-free streets, and similar solutions may be more effective if traffic authorities are tough enough to impose penalizing restrictions and if their locations are selected strategically so as to induce change in mode for the trip as a whole.[d] These are big ifs, and pricing strategies are more likely to succeed, as well as being more ideologically acceptable than the activities of local traffic commissars.

The desirability of actions to promote energy conservation is not dependent upon the existence or nonexistence of an energy crisis or shortage. A situation in which 5.6 percent of the world's population consume over 30 percent of the world's energy cannot represent an efficient allocation of resources, even from the narrow point of view of the country with the high consumption. The examples of waste are well known: the simultaneous running of air-conditioners and room heaters; the lack of automatic time switches on home central-heating systems; large cars with V-8 engines, automatic transmission, power brakes, power steering, and air-conditioning; the duplication of air routes with half-empty planes of different companies flying between two cities at the same hour; the all-night illumination of empty downtown offices; the profligate waste of

---

[d]The direct effects of banning cars may be quite small. For instance, it has been estimated that banning all cars and taxis from the central square mile of mid-Manhattan would reduce total vehicular traffic in the New York region by less than one percent.

energy in industry. If consumption continued to increase at recent rates (electricity consumption doubling every decade) the United States would need eight power plants in operation by the year 2000 for each one now—a new power plant every three weeks up to 1985 and one every week between 1985 and 2000. The amount of electricity produced over the next decade in the United States would probably equal the aggregate production since the first power plant was introduced in 1882.

Yet there is nothing inevitable about this growth (indeed, there has been a minor hiccough since 1973). The exponential growth path implied by the constant percentage rate of growth experienced up to now cannot last indefinitely; as Kenneth Boulding once said, only a madman or an economist would expect exponential growth to last forever. Several studies have shown that consumption could be cut back—by more efficient use, by measures to reduce demand, and by higher prices. High prices—particularly artificially high prices— have all kinds of nasty consequences, from regressive impacts on poor consumers to severe distortions to the allocation of resources. But most analysts agree that substantially higher energy prices could restrain demand. Professor Chapman of Cornell University, for example, has estimated that variations in the price of electricity and, to a lesser extent, in population growth could alter projected electricity demands in the year 2000 by 500 percent. A doubling of relative electricity prices would increase demand over the period 1970-2000 by a mere 33 percent. If electricity prices remained constant, however, relative to the general price level demand might increase by two-and-a-half times, and possibly by more. A RAND analysis estimated that increasing energy prices might reduce demand to one-half of the level projected by the electric power industry. A recent study of the residential demand for electricity by Halvorsen[10] suggests the following conclusions: the long-run price elasticity of demand is fairly high, at least unity; the cross-elasticity of demand with respect to gas price is significant but small; and the income elasticity of demand is less than one. If past trends in price and other variables continued, total residential electricity consumption in twenty years would be seven times the present level; if the real price of electricity remained constant, the ratio of consumption in twenty years to current consumption would be 2.9; if the real price of electricity increased by 50 percent, this ratio would only be 1.8. In 1974, for the first time for over twenty years, energy consumption fell. This was a year of rapidly rising energy prices, though other factors—such as the economic recession and mild weather— also had some influence.

These arguments and data are not deployed to press for a strategy of forcing up energy prices, but rather as an illustration of the thesis that demand growth is not unstoppable. A much more satisfactory means of adjusting prices than an overall artificial increase such as raising the federal tax rate per unit of electricity is to attempt to improve the structure of prices to reflect real costs more accurately. For instance, many utility companies operate promotional rate structures, which make high consumption rates cheaper in terms of average unit price. If the industry now functions under conditions of increasing costs and supply inelasticities, it would be more appropriate to charge higher marginal rates to bulk consumers. Another method of reforming the price structure, suggested by Walter Heller among others, is to include the environmental and social costs of producing energy in the price of electricity. This would be an improvement rather than a distortion of the price mechanism. True, the monetary quantification of some of these environmental costs is by no means an easy task, but even an approximation would represent a beneficial change in the societal allocation of resources.

However, the difficulties of attempting to measure these social and environmental costs should not be minimized. Some of the more obvious problems include the quantification of the monetary costs of pollution-induced illness (medical expenses and loss of work, not to mention the costs of pain and discomfort), property damage due to pollution, the impacts of thermal pollution on the fishing industry, visual blight of power plants and transmission lines, radiation exposure, damage to the health and safety of miners, amenity and recreational disruption of strip mining and other activities, and the separation of pollution impacts from different sources. These issues are not going to be resolved easily. In a second-best world it may be practical—disregarding the warnings of economic theory—to raise energy prices so as to reflect these social costs approximately, but to ensure that the increased revenues are used to control environmental impacts or to compensate for their effects, rather than simply accruing to the utility companies.

If such pricing strategies are regarded as running ahead of the political will of government at present, not to mention public opinion, a host of preliminary actions may be taken to induce energy conservation, even though some may take a long time to have much effect. These include measures to influence consumer behavior, to regulate the actions of the utility companies, and to induce design changes by manufacturers and construction firms. Most of them involve prohibitions, regulations, and controls, but not of a kind that could be regarded as major infringements with personal liberty. The possibilities include mandatory labeling of the amount of electricity used in various appliances (in fact, already widely adopted, without much noticeable effect); banning electric heating in the absence of major energy-saving design improvements (thus avoiding cases such as the infamous New York World Trade Center, which uses energy at the rate of 80,000 kilowatts per hour); total replacement of incandescent by much more efficient fluorescent lighting; limitations on billboard, advertising, and store lighting; stricter standards for building materials, heating systems, and insulation; tax incentives and investment subsidies to promote research into energy-saving appliances, machinery, and structures; controls on promotional advertising by electrical manufacturers to some degree and stricter controls on utility companies (some are taking a much more responsible attitude than in the past);

compulsory revisions in utility rate structures to eliminate the practice of offering lower unit prices as consumption increases and to impose surcharges on peak usage; and publicly financed educational campaigns to encourage conservation.

Even if all these measures could be agreed upon and introduced simultaneously, there is little reason to expect massive energy savings overnight. Design improvement and equipment and structure replacement have very long lead-times. Habit changes cannot be induced in the short run. The necessary modifications in life styles are not major, but may be difficult to achieve in less than a generation. The strongest hope for change is that attempts to persuade will be reinforced with economic incentives. Between 1946 and 1971, the price of electricity in constant prices fell year by year. Regardless of one's stand on the world energy situation, it would be unreasonable to expect consistent relative price falls over the 1971-96 period.

PIB made some estimates of the potential savings through energy conservation attainable by 1985. These are summarized in Table 9-2. To achieve these savings would require strong conservationist policies. For example, the 2.85 Q. Btu savings in transportation would involve a 20 mpg mandatory auto fuel economy standard (saving 1.9 Q.) and other disincentives to automobile use. Similarly, the savings in residential and commercial energy use would require a combination of incentives (tax credits for insulation and retrofitting, estimated to achieve economies of 0.8 Q. in homes and 0.2 Q. in commercial buildings) and compulsory standards (lighting regulations for commercial buildings, new building codes for both residences and commercial buildings, and household appliance efficiency controls). As suggested by the data of Table 9-2, the scope for savings varies among sectors—high in transportation and the residential and commercial sector, but relatively low in industry and the utilities. To achieve improvements in thermal efficiencies of electricity generation would require a reversal in the experience of recent years, and is dependent upon major technological changes and successful R and D, as indeed is energy use economy in industrial processes. In other words, these are estimates of *maximum* savings, which may not be achieved. Moreover, even savings on this scale would reduce energy demand by less than 10 percent—a substantial reduction, but unlikely to alter the broad pattern of energy trends over the next decade.

In the longer run, there are those who speculate on the likelihood of drastic changes in spatial organization, methods of transportation, social behavior, and choice of technology that conform to the principles of an energy-saving society. Here, the lead-times are tremendous. The mass transportation society and the return to a high-density pattern of city life would take so long to bring about that the hump in the energy problem should be over well before the desired changes had been achieved. The creation of a *safe* nuclear power industry or the harnessing of other forms of power require predominantly technological solutions, and these present many fewer obstacles than the social engineering remedies.

10 Energy and the Environment

Pollution Impacts

The trade-off between energy use and environmental protection is a critical question for society and its policy-makers, arising from the fact that the energy sector and fuel combustion are the biggest polluters of all. The transportation sector, for instance, is responsible for 75 percent of carbon monoxide, 57 percent of hydrocarbon, and over 50 percent of nitrogen oxide emissions. (Diesel engines would make a big difference, with only 2 percent of the carbon monoxide and 35 percent of the hydrocarbon emissions of gasoline, though much worse from the point of view of nitrogen oxides and particulates.) Fuel combustion from power stations generates over one-half of sulfur oxide emissions and a sizeable share of particulates (second only to industrial processes). Power stations are also responsible for substantial thermal water pollution from thermal discharges, while all types of mining create major water pollution problems for rivers and underground watercourses. Ocean pollution results from oil spills, though these are more frequent at offshore production facilities and terminals than from tankers in motion. Finally, the land is polluted as much as the air and water by energy use. Mining creates large volumes of solid wastes, and land disruption—from mining, especially surface mining, oil and gas fields, power transmission lines, and sites for power stations—is on a massive scale, almost 20 million acres in total.

Table 10-1 presents some data on the pollution impacts of energy use, present and future. The latter are speculative, since there are three major determinants of future pollution impacts: the severity of pollution controls and improvements in pollution abatement technology; the level of total fuel consumption; and changes in the fuel mix. The estimates in Table 10-1 refer to PIB's forecast for 1985 under "business-as-usual" conditions.[a] Also, they show the aggregate national impacts; regional differentials may vary widely as a result of changes in the fuel mix or even within a single fuel (for example, the choice between low-sulfur western and eastern coal, or the degree of replacement of mainland by Alaskan oil). Water pollution is expected to be drastically reduced, apart from thermal discharges. This prospect reflects the full implementation of the Federal Water Pollution Control Act of 1965 by 1983. The improvement in air pollution

---

[a]Assuming an $11 per barrel world oil price. If oil prices were lower, pollution levels would drop despite higher fuel consumption. This is because more oil imports would be substituted for coal.

Table 10-1
Pollution and Energy Use, 1972 and 1985

	1972	1985
Water		
Dissolved solids (tons/day)	37,000	5,800
Suspended solids (tons/day)	7,600	300
Thermal discharges (billion Btu/day)	19,500	24,000
Oil spills (barrels/day)	368.3	170.6
Air		
Particulates (tons/day)	1,800	2,300
Nitrogen oxides (tons/day)	38,000	46,700
Sulfur oxides (tons/day)	58,900	53,700
Hydrocarbons (tons/day)	33,200	18,800
Carbon monoxide (tons/day)	7,900	1,400
Solid Waste (1000 tons/day)	900	1,100
Land disruption (1000 acres)	19,800	26,700

Source: Federal Energy Administration, *Project Independence Blueprint* (Washington, D.C.: U.S. Government Printing Office, 1974), pp. 33, 225, 227.

is unlikely to be as marked, partly because of increased reliance on coal, and levels of nitrogen oxides and particulates might in fact rise. Solid waste residuals will increase, especially if oil shale developments go ahead rapidly. Land disruption will increase substantially, mainly because of extensive surface mining, and to a lesser extent because of oil shale and hydroelectric power developments. However, these estimates should be treated with discrimination, since pollution loadings are sensitive to the type of energy strategy adopted and to major changes in environmental standards.

Environmental Implications of Power Plants[1]

Power plants are the dominant source of environmental hazards in the energy sector for several reasons: the high, and increasing, share of electricity in total energy supply; the scarcity of relatively safe fuels (natural gas, oil, and low-sulfur coal) for power stations, leaving coal and nuclear power as the expanding fuels; the ubiquity of power plants; and the failure hitherto to solve their major environmental problems. Coal-burning power plants are regarded as being the most dangerous environmentally, but at least the dangers are well known—even if it is difficult to determine the most appropriate environmental standards. The problem with nuclear power plants is that the environmental risks are much more difficult to calculate, and the solution for some hazards is unknown. These include the management of highly radioactive wastes, the risk of a major reactor accident, radiation spills during the transport of nuclear fuels and wastes, and—where cooling by water is used—the effects of thermal pollution.

Nationally, electric power plants discharge about one-half of all sulfur oxides, one-quarter of all nitrogen oxides, and one-quarter of all particulate matter. In some areas their proportionate impacts are even greater; in New York state, for example, power plants account for almost three-quarters of sulfur oxide emissions, and well over a third of nitrogen oxides. Some estimates suggest that sulfur oxides and particulates may account for the premature death of up to 2000 New Yorkers each year. Annual average SO_2 levels are three times and nitrogen oxides double the federal government's acceptable standard.[2] Although automobiles discharge about three-fifths of all air pollutants by weight, they make little contribution to sulfur oxide and particulate pollution, and their role in nitrogen oxide emissions depends on many local factors. At present, no satisfactory (i.e. inexpensive) technology for controlling the major power plant emissions—especially SO_2—has been developed, not to mention applied, and without a breakthrough in this respect there will be several-fold increases in emission levels as the utilities respond to meet rising power demands. A single plant of 875 Mw capacity burning one percent sulfur fuel oil might emit about 55 million pounds of sulfur oxides, 19 million pounds of nitrogen oxides, and 400,000 pounds of particulates each year.

Although there is debate about the precise critical levels—for instance, some argue that the federal standard for sulfur oxide emissions is arbitrary, the health risks of air pollution are unanimously accepted. Historically, excess deaths have been attributed to SO_2 and particulates in several cases: the Meuse Valley, Belgium in 1930, Donora, Pennsylvania in 1948, London in 1952, 1956, 1959, and 1962, and New York City in 1953, 1962, and 1966 are among the most notable examples. Inhalation of SO_2 may be associated with chronic respiratory diseases such as emphysema, bronchitis, and bronchial asthma. Possibly more serious, oxidization of SO_2 to form sulfur trioxide, which then combines with water to form sulfuric acid, can irritate eyes, nasal passages, the throat, and the lungs, and can result in lung damage. In addition, it has corrosive effects on inanimate materials. Particulates have similar effects. The utilities' claim that almost 99 percent of the weight of particulates is removed by control apparatus is hardly comforting, since the smaller particles generate the greatest biological damage by penetrating the lower respiratory passages. Nitrogen dioxide is the most dangerous of the nitrogen oxides, inducing respiratory diseases and cellular disintegration as well as minor effects in terms of eye, nasal, and other irritations. The adverse health effects level of NO_2 emissions is exceeded in 85 percent of large cities (>500,000). Nitrogen oxides react under sunlight to produce hydrocarbons (photochemical smog), and power plants are a significant contributor to this problem, though less important than automobiles and some industrial processes. Carbon dioxide is also emitted in quantity by fossil-fuel power plants, and has been identified as a major factor in the overheating of the earth's atmosphere. The consequences are shrouded in uncertainty. One estimate[3] suggested that the doubling of atmospheric CO_2 would be needed to raise average temperature by $2°$ Centigrade, though this could result in melting ice caps. Nordhaus[4] estimated that energy consumption might increase by the year 2030 enough to raise CO_2 concentrations by over 40 percent, serious but probably not dangerous. Others hold a gloomier view.

There are only two strategies potentially available to control air pollution from fossil-fuel power plants: technologies to prevent emissions into the atmosphere and regulation of the type of fuel used, particularly its sulfur and ash content. *Commercially* proven methods for controlling sulfur oxide emissions have yet to be introduced. Techniques for dealing with nitrogen oxide by altering the combustion process within the boiler are known, but are being used by only a few utilities, particularly in California. One reason for the delay has been the low levels of federal and nonfederal research spending on stack removal and sulfur recovery processes—perhaps only about 10 percent of the effort required to yield solutions within a 5-10-year period. The failure to recover sulfur is wasteful in the sense that annual consumption of mined sulfur is about 16 million tons, while the equivalent of more than 12 million tons—and the amount is rising steadily—is emitted each year. In view of the increasing demand for power and the slow rate of progress in solving these problems, emissions are likely to increase in spite of the transition to nuclear power, the substitution of gas and low sulfur fuel, improved techniques of combustion, and technical advances in stack-gas treatment and in sulfur recovery.

Despite the abundance of fossil fuels in the United States, low-sulfur fuels are in short supply. High-sulfur coal reserves are much more plentiful than low-sulfur reserves, supply constraints in natural gas are severe, and imports of liquefied natural gas can play only a marginal role in consumption (optimistically, perhaps 5 percent of consumption by 1980), and there is a history of government-promoted output restrictions in the domestic oil extraction industry which have been more in the interests of the oil producers than the public. Even if low-sulfur fuel regulations were improved, it is arguable that SO_2 emissions would still be uncomfortably high. The only sound method of meeting standards would be large-scale conversion to natural gas,[b] but this is infeasible because of insufficient supplies and reserves. Indeed, the more likely development in the current situation is that federal standards for emissions will be relaxed in order to permit more burning of coal. The weakness in the case for maintaining present

---

[b]At first sight, hydroelectric power looks attractive. However, most potential undeveloped sites are at remote locations where there is little local demand and transmission costs would be high. Moreover, despite the absence of the main forms of power plant pollution, there are social costs associated with flooding scenic areas, the value of which increases over time because of their inelastic supply (J.V. Krutilla, "Conservation Reconsidered," *American Economic Review*, 57 (1967), pp. 777-86). Also, there is a nitrogen supersaturation problem in the water flowing over the spillways of hydroelectric dams, which has adverse effects on fish population.

standards is that the epidemiological evidence—though clear in identifying the health risks associated with sulfur oxide emissions—is not sufficiently sensitive to pinpoint the critical levels.

Although nuclear power plants avoid most of the air pollution problems associated with the burning of fossil fuels, they create other environmental problems which may be more intractable and are certainly less understood. The rate of substitution of nuclear for fossil-fuel plants is uncertain, depending on relative costs, production bottlenecks, and the decision-making processes of utilities, but the AEC's *most likely* forecast is 280,000 Mw in 1985, 508,000 in 1990, and 1,200,000 in the year 2000—and the latter figure implies 900-1000 large individual plants. The expected rate of growth has serious implications for reactor accident risks, waste disposal, and other nuclear environmental problems.

The AEC's record is far from satisfactory. Prior to October 1970, it was responsible both for promoting nuclear power and for the regulation of plants. From that date the Environmental Protection Agency was made responsible for establishing radiation standards and radioactive emission limits, though the AEC remained the enforcement agency and retained control over siting and safety, and these functions have now been taken over by the new Nuclear Regulatory Commission. The AEC has frequently failed to keep the public fully informed about the risks of nuclear power reactor accidents. It has acted on several occasions as if public relations were more important than knowledge about safety, and as if it represented the vested interests of a private industry rather than the public good. Its research budget on the safety of the type of reactors that will dominate the next decade or more was cut in favor of development of the more distant breeder technology. The AEC has supported the federal Price-Anderson Act of 1957, which limits the financial responsibility of the utility companies and their suppliers for radiation damage resulting from major reactor accidents. Finally, the criteria used in siting plants are vague. The claim is that plants are located far away from population, but—to mention one example—the Indian Point plant in New York state is only one mile away from an area with 25,000 people. The development of large nuclear complexes may resolve this problem if isolated sites are carefully selected. However, the risks of a much larger nuclear accident may be greater and more costly even if the general population is far away.

The environmental problems associated with a nuclear-power-based economy are difficult to evaluate because of ignorance about the existence of a threshold limit which avoids biological risk. Although breeder reactors may offer conservation advantages that offset environmental impacts, these compensating advantages are not shared by the light water burners that characterize current technological practice. The distinctive feature of nuclear plants in terms of cost is much higher capital costs but anticipated lower operating costs. The utility rate structure and method of calculating returns tend to favor nuclear power, because the high capital costs of the plant can be included in the utility's rate base, thereby increasing its allowable revenues.

The nuclear reactor safety record hitherto has been quite good. There have been some notable minor accidents: a plutonium production reactor malfunction in England in 1957; the mismanagement of a single control rod in an experimental AEC reactor in Idaho in 1961, killing three men; a fuel meltdown in the Enrico Fermi power plant in 1966. The nuclear safety statistics read very well, but it must be remembered that they are based on the performance of a score or so modest size reactors. If a major accident were to happen, the consequences could be very serious. Many years ago (in 1957) the AEC estimated the *worst* that might happen to a small (200 Mw) reactor that had been operating for six months thirty miles from a city of one million on the assumption that all safeguards failed and evacuation of the residents could not be carried out. If 50 percent of the radioactive products were released, became airborne and were dispersed, the death toll might be 3400, the casualty list 43,000 and the property damage, $7 billion (at current prices). Deaths would occur up to fifteen miles, with injuries up to forty-five miles.

By 1990 most sites will have power plants in the 1000-4000 Mw range, with some as large as 6000 Mw. Assuming 1000 reactors by the year 2000, much depends on the forecast odds. The official view is that the accident risk is 10,000: 1 per reactor-year; that is, one accident per century if there were 100 reactors, or one accident per decade with 1000 reactors. Of course, safety precautions and postaccident actions might reduce the consequences of an accident appreciably. Nevertheless, the odds are not as strong as the crude ratio suggests, once account is taken of the number of reactors that might be operating. Also, some observers assess the risks as being ten times larger than the official evaluation. Furthermore, the risks of explosive accidents are higher in the case of breeders than for light water reactors. Although normal radioactive emissions are only a tiny proportion of the declared "safe" standards, the performance of nuclear fuel reprocessing plants is inferior. For instance, the reprocessing plant in West Valley, New York had effluents up to 18 percent of maximum allowable concentrations, even when operating at less than 50 percent of capacity.

The thermal pollution effects of power stations are serious, particularly since 80 percent of thermal pollution arises from electricity generation, and nuclear plants use 50 percent more water than fossil-fuel plants of equivalent size. The water requirements of water cooling are huge. Given the very high proportion of steam-electric plants in total electricity generation up to the 1990s (perhaps 85 percent), if water cooling were to continue unchecked, one-half of the average daily natural runoff of water in mainland USA would be needed for cooling by the year 2000. Also, the effects on aquatic life and water quality could be serious; there have been at least two major fish kills at Indian Point on the Hudson River since operations began in 1963. The alternative to water cooling is the use of cooling towers, widely adopted in Europe. This is a much more expensive procedure than water cooling. Wet cooling is unsuitable for Northeast locations because of climatic conditions—high humidity and low temperatures. Dry cooling is even more costly and probably less effective, and its environmental effects are unknown. However, this is the direction in which the industry must move if water-flow cooling is to be controlled.

The most effective solution to radioactive waste disposal has yet to be determined. The AEC practice of permitting the storage of high-level waste at the fuel reprocessing site for up to ten years after separation from used fuel is based on economics rather than safety. The most feasible prospect is to solidify high-level wastes and to store them in salt mines. In 1972 the AEC purchased a 1000-acre site near Lyons, Kansas, for $3.5 million, as a demonstration project (total estimated cost $25 million) for a salt-mine depository. The wastes would be packed in sealed steel containers, each ten feet long and containing more than 1 million curies of radioactive material, and set in holes in the floors of rooms in the salt formations about 1000 feet below the surface. Radioactive decay would generate heat, raising the temperature up to more than 200°F., but the plastic nature of the salt would prevent this from becoming a major problem. However, the project has since been abandoned, because subsequent testing found it geologically unsuitable and the mines contained water.

The land demands of the electric-power industry are considerable, both for sites for power plants and for the rights-of-way used for power transmission lines. At present, there are more than 300,000 miles of electric power transmission lines in service in the United States, occupying more than four million acres of land. Less than one percent of this length is underground, and there has been little interest in research into underground transmission techniques such as the use of cryogenic (superconducting and supercooled) cables. By the year 2000 it is estimated that—in the absence of a shift to underground cables—almost 11,000 square miles would be taken up with electric transmission lines, an area equivalent to that of Massachusetts, Delaware, and Rhode Island combined. The effects of a transmission system on this scale on scenic beauty, property values, radio interference, and safety are substantial.

The power plant site demand problem is also far from negligible. Hitherto, most power plants have been located on riverside or coastal sites. For instance, California has established 85 percent of its power plants on tidal waters. The conflict between power plant requirements and the wildlife and recreational preservation of coastal resources creates a familiar problem. One solution is to enforce air rather than water-flow cooling. Another, more tentative remedy is to build power plants offshore, using shipyard technology.[5] There is a choice between floating platforms moored to the ocean floor (submersible stations, ship hulls, or barges) and fixed structures (manmade islands, fixed-pile platforms, jack-up platforms, or grounded barges).

More generally, it is important to minimize the site demands of power plants and to locate new plants, whether fossil-fuel or nuclear, at a distance from population centers. The space requirements of a 3000 Mw generating station, including the land needed for fuel storage, heavily favor oil (150-350 acres) or nuclear (200-400 acres) rather than coal (900-1200 acres). In many states power site decisions have not been subject to a sufficiently strong degree of control in the public interest. The Public Service Commission of New York, for example, has allowed huge expenditures on land for future power plant use to be included in the base rate. This encouragement of utility companies to buy up land is difficult to reconcile with environmental planning considerations.

Although evaluation of optimal levels of expenditure on R and D is difficult (see pp. 167-72), if not impossible, there is prima facie evidence that the utility companies have neglected research. Research expenditures are about one-quarter of one percent of gross revenues, or about one-eighth of spending on advertising and sales promotion. One method of promoting expenditures on research might be to allow utility companies to count R and D expenditures in its rate base. Two examples of research neglect have already been mentioned: techniques for controlling emissions and the development of low-sulfur fuels. A pretext used to justify this is that the utility companies live in expectation of a relaxation of environmental standards, and hence leave most of the research in these areas to the government. This weak argument does not apply in a third area—the low thermal efficiency of power plants. About 80 percent of electricity supply in the United States is produced in steam-electric plants by a thermodynamic process known as the Rankine Cycle. This wastes almost two-thirds of the heat produced, as well as generating large quantities of air pollutants. After 1966 (and up to the energy crisis of 1973) energy consumption per GNP dollar rose, a reversal of fifty years of steady decline. A major factor in this change of trend was the limited scope for further technical improvements in electricity generation in the absence of major shifts in technology.

There is scope for research and development in many directions: combined-cycle units, fuel cells, and magnetohydrodynamics (MHD). MHD, for instance, is a method of converting heat directly into electrical energy without the need for a conventional turbine and generators. It could result in thermal efficiencies of up to 60 percent. However, research expenditures on MHD have been meager, despite an estimated requirement of $500 million over a 15-25-year period. It is possible, of course, that this particular process may be too expensive to develop, and not necessarily the best of the alternatives potentially available. Yet its neglect is indicative of the attitude of the industry to research.

Coal and the Environment

Suppose the basic arguments of the pessimists are accepted. Then, one of the major problems facing the United States is the probability of an energy gap, occurring at some date between the present and the end of the century. The reason for this gap would be supply inelasticities in oil and natural gas prior to the development of the nuclear-powered economy or transition of some other nonfossil-fuel technology. Assuming a failure to develop alternative modes of transportation not reliant on oil, a substantial share of oil supplies would have to be reserved for automobiles and planes. How could total energy demands be met under these conditions without a lower level of GNP, drastic restrictions in domestic energy use, and gasoline rationing? The missing element in this scenario is coal. Estimates of coal reserves are speculative because the methods used involve rough calculations about the depth and amount of coal lying below a given area of land in regions where coal is known to exist. Also, there is the usual problem—perhaps more extreme in the case of coal—of the distinction between total reserves and recoverable reserves at feasible price levels and with current or foreseeable technology. Nevertheless, regardless of the estimates chosen, few would argue that potentially mineable coal is equal to several hundred years of consumption at current rates. US proven reserves exceed 430 billion tons, and total reserves are probably considerably higher than three trillion tons. These are substantial when compared with forecasts of production, since few predict a coal output that would exceed 1.5 billion tons in the late 1980s; much more than double current output, yet only a minute fraction of proven reserves.

These data imply that the energy gap is a fiction. What, therefore, is the problem? As S. David Freeman says, "There are two things wrong with coal today, we can't mine it and we can't burn it." Coal is regarded in many quarters as an environmentally unacceptable fuel: deep mining involves serious health and safety risks, strip mining disturbs visual amenity, conflicts with leisure and recreational objectives, and can lead to ecological damage, and the burning of coal is a major source of air pollution. From this point of view, coal is a demand-constrained rather than a supply-constrained fuel. Apart from the dangers of deep mining, it is probable that the environmental problems of using coal can be solved, but not immediately. The conflict between energy needs and environmental goals may be critical in this case. How should these two important objectives be traded off against each other? The need for trade-off will be affected, of course, by how serious the supply shortages of other fuels are. Since two-thirds of coal consumption occurs in power plants, much depends on what happens in the electric power industry. The risks of future interruption in supplies of oil and natural gas will tend to discourage decisions in favor of power plants running on these fuels in the near future. The choice is likely to be between coal and nuclear power. In view of the escalating capital costs of nuclear power plant construction and the rising price of uranium, the balance might swing heavily in favor of coal, were it not for the environmental repercussions. The question of whether or not the environmental constraints imposed are appropriate is critical.

The coal lobby among energy economists believe that the pendulum has swung too far in favor of environmental controls, to the detriment of the economy and general social welfare. They start from the proposition that it is unfortunate if a fuel that accounts for 90 percent of world fossil fuel reserves is only going to meet about 20 percent of US energy needs by the end of the century. Unless the environmental demand constraints on the use of coal are enforced, they argue that the real situation in US energy is one of excess supply. As for coal, its supply elasticity is very high. Strip mining is highly profitable, even more so with recent energy price increases, and choosing the most appropriate method of extraction results in easy compatibility among the goals of increasing supplies at reasonable cost, profitability, and restoration of the landscape to its premining state. The supply of low-sulfur coal, mainly from western fields, is limited in the short run, and in the longer run if environmental problems are not solved. High-sulfur coal is much more plentiful, and could be used without desulfurization (though this may be practical and not too costly in the medium term), provided that environmental sulfur oxide emissions were set no higher than 2.1 particles per million. Such a relaxation would allow supply to increase to fill any likely demand-supply gap before the large-scale introduction of nuclear power. The argument is further based on the hypothesis that there is no convincing epidemiological evidence to show that the differential between present standards and 2.1 p.p.m. has any additional ill effects on health, and that the absolute effects are small compared with other, possibly preventable, social costs such as road accidents. The advantages claimed for this solution are considerable: easy solution of the energy problem; buying time for research into sulfur oxide control technologies and cheaper desulfurization processes and into alternative sources of power; avoiding crash programs of research into coal gasification and liquefaction which may yield only high-cost solutions that could be undercut by a drop in the world oil price; and no need for draconian measures of energy conservation.

This analysis is not without merit, but it is overstated. It underplays the environmental impacts of strip mining, and it underestimates *total* restoration costs and their enforcement. It assumes that in deep mining a balanced course can be steered between manpower shortages on the one hand and uneconomic rising wage costs on the other. It ignores the fact that power plant pollution costs cannot be avoided by individuals in a world of less than perfect mobility, so that we should be especially careful about subjecting populations to increased emissions. At this critical stage in the history of environmental controls, official relaxations of emission standards have damaging psychological effects by discouraging research, both in the public and private sectors, into emission control technologies. Also, in the absence of clear-cut epidemiological evidence, the onus of proof should rest with the polluters rather than the polluted.

Perhaps the mistake in the analysis is to argue for a *general* relaxation of standards. It may be more reasonable to argue for variation in standards among areas according to some formula that takes into account population size, density, climate, other pollution sources, and other factors affecting concentration in particular areas. Relaxations could be granted on an ad hoc basis in the less risky areas for a prescribed time period, so as to continue to encourage pollution control research. Also, if research into alternative fuels can be wasted by cutting of the world oil price (always feasible in view of the wide gap between selling prices and Middle East production costs), similarly, investment in coal mines and in coal-fired power plants may equally be undercut. Of course, since the adaptation of coal power stations to oil or natural gas is easy and not too costly, this objection should not be exaggerated.

With adequate safeguards, therefore, the switch to greater use of coal is worth trying. However, such a strategy should not be interpreted as a green light for environmental destruction. But the level and enforcement of environmental regulations will be the primary determinant of the fuel supply situation over the next decade.[6]

Stronger measures are needed to protect the landscape against strip mining; at the time of writing, the struggle for effective federal strip-mining legislation continues. Even by 1970 the US Geological Survey estimated that 1.3 million acres had been damaged by strip mining, and although the National Coal Association estimated that about 81,000 acres had been redeveloped or reclaimed during 1971, this is but a token effort relative to the scale of the problem. Even with rising reclamation costs or higher reclamation standards, the additional costs with either of the two main techniques (modified block-cut or contour backfilling) would not raise the price of coal beyond competition with other fuels. Slope limitations (e.g. prohibiting surface and strip mining on slopes greater than 15 or 20 degrees) would help minimize environmental disruption, but would rule out obtaining much low-sulfur coal.

The environmental struggle that could accompany an attempt to exploit western coal in an all-out fashion has yet to be fought. The conflict between preservation of the countryside, landscape and amenity resources, and economic objectives would be much sharper in the case of the West than in the old-established industrial and coal-mining areas of the East, which grew and prospered on the basis of coal and steel. The conflict would be accentuated by the increasing competitiveness of western coal in an area such as Chicago, or the Midwest generally. Western coal has the advantage of low sulfur content but offsetting drawbacks of lower grades and higher transport costs. Nevertheless, it becomes cheaper than eastern coal when account is taken of rising wages in underground mining and the scrubber costs of treating plants using high-sulfur coal.

In view of the short-term scarcity of low-sulfur coal, another interim arrangement would be to make more use of distillate and residual fuel oils by a variety of methods: blending, burning crude, desulfurization, and increased imports. The first three solutions have different environmental impacts, while the last raises other problems such as supply difficulties and adding to the balance-of-payments deficit. Nevertheless, a mix of these measures could meet any shortfall in supply over the next few years.

Automobile Emissions

Automobiles are a major source of carbon monoxide, hydrocarbons, nitrogen oxide, and secondary pollutants in the form of photochemical smog in urban areas. Since 1968 attempts have been made to tighten up on automobile air emissions. Where fuel conservation is also a policy objective trade-offs arise between controlling pollutants and achieving fuel economy, particularly since the catalytic converter introduced on 1975 models with the aim of achieving both goals has proved to be a premature solution because it emits damaging sulfate particles. Actions to reduce the use of the automobile, either by disincentives or by providing alternative means of transportation, have the advantage that they reduce pollutants and save fuel, but hitherto have had little success. Measures to stimulate a switch in tastes in favor of small cars would also help to promote both objectives, and perhaps in a more palatable way. However, in 1975 the administration relaxed its emission standards in return for a commitment from the auto industry that it would press ahead with attempts to achieve a 40 percent fuel economy improvement target by 1980. In the medium term, perhaps a decade, the problem of automobile emissions should be amenable to a complete technological solution based on a stratified charge, diesel, or some other kind of engine.

Oil Spills

The problem of oil spills could become more serious with the switch in location of oil production from onshore to offshore developments—off Alaska, the Pacific, the Atlantic,[7] and the deeper parts of the Gulf of Mexico. Since the probability of oil spills is influenced by climatic conditions, southern Alaska could be a particularly vulnerable area with its winter storms, earthquake risks, and tsunamis (high-intensity ocean waves). Also, the Gulf of Alaska area is a major flyway for more than 200 species of bird, and the fisheries are important for pink salmon, perch, and shrimp. Bird and marine life risks arise in the other offshore regions too.

The full ecological risks of oil pollution are still unknown. Bird populations can be endangered by oil coating which reduces insulation and by ingesting of oil during preening. There are cases where many birds have died: between 6000 and 15,000 died as a result of the Santa Barbara spill of 1969, for example. Fish populations are damaged by the death of eggs and larvae from exposure to hydrocarbon concentrations. Fully grown fish can avoid contaminated areas, but shellfish cannot. Salt marshes and estuaries can be severely affected by chronic oil spills. The costs of cleaning up beaches if oil spills come ashore may be very high, especially in recreational areas. Although the large spills (such as the *Torrey Canyon* incident off the English coast in 1967 and the Santa Barbara spill of 1969) receive most of the publicity, the accumulation of smaller, chronic spills over a long period of time may do more damage. More research needs to be undertaken on the resilience of marine and coastal biological systems.

No matter what protection procedures are built into oil production and transportation, some spills are inevitable. Greater care and safety can be obtained, particularly in offshore production, where over one-half of the spills related to the oil industry occur. There has been much controversy about the types of tankers being built, and about whether they are strong enough to withstand the rough storms of the Gulf of Alaska and the North Pacific. However, only twenty-two spills in 1971 and thirty-two in 1972 were from ships out at sea (as opposed to spills at the terminals), though in 1972 some of the spills were quite large.

The Trans-Alaska Pipeline Debate

Apart from the extreme fringes of the environmentalist lobby, few would argue that environmental protection has an infinite value and hence is sacrosanct. Any decision that involves environmental disruption should be subject to scrutiny on the basis of a comparison of benefits and costs, and there is no reason why—in view of the estimation difficulties—the measurement of environmental costs should not err on the high side. Moreover, there are usually alternative ways of achieving a given end, and if environmental protection objectives rank high among societal preferences (however expressed), an efficient solution may well be one that holds the level of disruption down even if it diverges considerably from that which would maximize the net present value of *private* benefits minus costs. It should be the solution which maximizes net social benefits, provided that societal preferences are accurately reflected in the value attached to intangible environmental costs and benefits. The case of the Trans-Alaska Pipeline (TAP) is an interesting illustration of these points.[8]

In 1968-70 exploration resulted in the discovery of a major oil field on the North Slope of Alaska—the most important single discovery in the history of the domestic crude oil industry. Several methods were possible for transporting the oil to market. These were variants of three basic strategies: using icebreaking tankers to deliver the oil by water all the way; an overland pipeline through Canada; and a combined land-water system involving a pipeline to a southern Alaskan port and then shipment by conventional tanker. The first was rejected on technical grounds, and an oil company consortium chose the third method. They proposed building a pipeline 789 miles long between Prudhoe Bay and Valdez and a road 373 miles long from Prudhoe to the Yukon River, where it would link up with the existing road.

About the same time the National Environmental Policy Act of 1969 was passed. This required federal government agencies to file an environmental impact statement whenever their actions were expected to affect the environment. This gave an opportunity to certain environmental groups to take legal action on the grounds of impacts on the caribou herds of the North Slope; the possibility of pipeline breaks due to melting permafrost, earthquakes, or avalanches; destruction of the tundra and its ecology; marine pollution due to tanker loading operations; and the possibility of tanker accidents. Their case was reinforced by a different set of arguments; land claims by Alaskan natives and protests by Cordova fishermen of Prince William Sound about disturbance to their livelihood. These actions have failed to stop the pipeline. Cicchetti[9] compares the TAP proposal with four alternative routes for a Trans-Canadian pipeline (TCP-I to IV) on the basis of the following environmental considerations: permafrost, seismic risk, marine spills, physiographic barriers, effect on wildlife migration, and meteorological conditions. He found that an overland route through Canada to the Midwest was environmentally superior, mainly by avoiding the threat of earthquakes and avalanches in southern Alaska and not creating marine pollution. Such a route—especially the one from Prudhoe to Fairbanks, then using the Alcan Highway right-of-way to Edmonton, where it would link up with the Interprovincial Pipeline which carries Canadian oil to eastern Canada and the US Midwest—would have economic advantages as well.

The quantitative estimates of his study have been superseded by events, particularly by the rise in the price of world oil and by the greater need to preserve US domestic supplies for home use. However, these factors only reinforce his conclusion.

The basic finding, subject to the important qualification that no superior transportation and marketing alternatives exist, was that the present economic value of TAP to the nation was very high—at the time $3-$6 billion assuming a 10 percent discount rate, now very much higher—and that the environmental damage would have to be at least equal to the economic value to justify abandonment of North Slope oil. Although it is unlikely that the environmental costs were high enough to justify a "no development" decision, the Department of the Interior's environmental impact statement submitted in 1972 frankly admitted that the environmental costs were real enough and considerable.[c]

TAP is not only an unsatisfactory choice environmentally, but it is not the most profitable transportation system for North Slope oil. A TCP supplying the markets of the Midwest and the East Coast, where prices are higher than on the West Coast, would be preferable for the nation, for the state of Alaska, and for consumers. In a free competitive market it would also be a more profitable route for the oil producers, because it would enable them to sell oil at higher prices without a commensurate increase in costs. To explain their preference for TAP it is necessary to take account of institutional and legal arrangements.

---

[c]The risks of environmental damage (such as melting the permafrost or interfering with the migration of the caribou) are being minimized by burying the pipeline underground or by raising it high above the land surface.

For example, the oil producers originally planned to take advantage of a plan to export excess supplies to Japan in foreign tankers and to import an equivalent amount of oil—again in foreign tankers—to the East Coast. This would avoid infringement of the import quota scheme and of the Jones Act, which requires higher-cost US ships to be used for trade between domestic ports, and by selling oil to themselves or exchanging exports for imports at the world price the oil companies would be able to reduce the wellhead price and consequently pay less royalties to Alaska (changing prices and supply conditions have destroyed much of the rationale of this Machiavellian scheme). Another, similar plan involved shipment to Central America in foreign tankers, a pipeline to the Caribbean, tanker transportation for refining in the Virgin Islands, and shipment to the East Coast. This circuitous route would be more profitable to the oil producers than an overland pipeline through Canada.

Other factors relate to the timing of production. Given the low tax rate on oil, the costs of delay in terms of foregone profits were high. British Petroleum (BP), one of the consortium, had an agreement to achieve a certain production target by 1977. Also, a TCP might involve sharing throughput with Canadian oil. The TCP scheme might also have led to delays because of negotiations with the Canadian government. TAP was undoubtedly the fastest route to develop and it suited the oil producers' interests—but at the expense of consumers, taxpayers, the US shipping industry, and the state of Alaska. Yet it was the solution permitted.

Everyone knows that the environmental impacts of production and consumption of energy are considerable. In some cases decisions have long been made, and there is very little that can be done about it. The costs have to be tolerated or the activities stopped. In other cases, such as emission standards for motor vehicles, it is possible to introduce legislation and controls to lessen the harmful effects of energy pollution. In a few cases, of which how to transport Alaskan oil is one example, the environmental problems are known in advance, and it may be possible to achieve stated objectives by choosing from alternatives a method that minimizes environmental disruption at minimum, or even negative, cost. Hopefully, in an environmentally conscious society corporations will begin to take decisions that recognize their social responsibility and consider the wider implications of their activities than the mere maximization of dollar profit. If the private sector fails to accept these obligations, then the government—as the guardian of social interests—must compel it. If Cicchetti's analysis is correct—and it appears convincing—the tragedy of TAP is that the industry lacked the maturity and the federal government lacked the political will.

11 Some Policy Considerations

Energy R and D

Although reasonable minds may differ about the role of technology in finding solutions to current problems, few would deny that in the energy sector there are some critical technological bottlenecks that, if and when surmounted, could radically transform the energy situation. Given the complexities of the modern world, new technology does not fall as manna from heaven but is the uncertain output of investment of capital and other resource inputs. Energy R and D expenditures are consequently an important element of energy policy. It is also hardly surprising that the traumatic events of 1973-74 should have led to demands that the government should spend its way out of the risk of another energy crisis by making the R and D investments which would develop alternative sources of energy to imported oil. If a higher degree of self-sufficiency is an acceptable energy goal, it is reasonable that one of several instruments to achieve this is public sponsorship, funding, or participation in energy R and D.

The difficulties begin when this argument is extended to deciding the optimal level of energy R and D. Theoretically, the answer is simple. Since R and D expenditures are a form of investment—a sacrifice of present for future consumption—projects should be funded up to the point where expected marginal social benefits and costs are equal. In practice, this is not very helpful, since the information is not available to measure these social costs and benefits. Interdependence of energy R and D projects, discounting of benefit and cost streams over time, and allowing for the considerable risks are additional complications. A simple way out would be to leave energy R and D to the private sector. Private firms will operate so as to equate marginal private benefits and costs. This would lead to an optimal solution only if all costs and benefits were internalized in the R and D undertaker.

Several reasons explain why this condition does not hold: the existence of technological spillovers and the gains from diffusion of knowledge; the benefits of public goods such as a clean environment; the distortions to competition due to the fact that most energy industries are closely regulated for antitrust, environmental, and other reasons; national security reasons for domestic supply dominance; the impact of energy on the economy as a whole; monopoly elements in particular energy sectors; balance-of-payments considerations (probably spurious, since the same argument could be applied to any other sector with a substantial import component); and the need to develop strategies for counteracting the monopoly power of OPEC.

These considerations suggest that the government has some role in energy R and D. Critical questions then become: How should the government intervene? How important is energy R and D compared with other energy policy instruments? The predominant form of government intervention is direct funding. However, this has limitations and leads to inefficiencies: distorted motivations; inflexibility (e.g. reluctance to abandon unpromising lines of research); the scarcity of alternative sponsors when research decision-making is highly centralized; bias in project selection (e.g. vulnerability to the pressures of special-interest groups); discouragement of private sector R and D; and the instability of government R and D funding (shown in the differential fluctuations of private and public R and D expenditures over the last two decades).

There is something to be said in favor of the alternative strategy that government should provide the conditions for the stimulation of private R and D spending in the energy fields. One attractive idea is joint sponsorship of projects, such as the agreement in July 1973 between the AEC, the TVA, and the Commonwealth Edison Co. of Chicago to build a demonstration plant for the liquid metal fast breeder reactor. Other possibilities include amendments in legislation relating to antitrust actions, patent laws, taxation, regulation, protection, and government procurement. Although various arguments can be used to justify action in each of these spheres,[1] the trouble is that very often they provide a cushion of protection or easy profits rather than a guarantee of substantial R and D activity.

Energy R and D has several limitations as a means of achieving energy goals. It is most appropriate to deal with long-range goals, such as combatting the effects of a gradual trend toward rising energy costs related to the depletion of certain energy resources, or reducing dependence on foreign supplies *in the long run*, or providing eventual solutions to environmental disruption problems. It is not an instrument capable of dealing with short-run energy-policy objectives. Hence, it has limited scope in yielding solutions to the difficulties arising from the recent energy crisis or to self-sufficiency targets relating to the year 1980. A little reflection makes this clear. Take a more efficient automobile engine, for example. Design and development, retooling of assembly plants *plus* replacing the existing stock of cars at current replacement rates, would take 15-20 years. Similarly, coal gasification—on the most optimistic forecast—could not be expected to contribute much more than 3 percent of gas supply by 1985. Shorter-term energy objectives may be met by other strategies, such as high excise taxes on energy consumption, mandatory conservation measures, even stimuli to more oil and gas exploration—since even the last of these has a shorter lead-time than R and D.[a] The upshot is that energy R and D can be determined

---

[a]If social science R and D has any practical utility, this is one type of energy R and D that could generate quicker results. Examples include devising strategies for combatting the international oil cartel, the design of feasible instruments for controlling demand or stimulating supply, development of optimal stockpiling schemes, and so on.

rationally only in the wider framework of energy policy as a whole. There may be less expensive and more effective ways of achieving goals than more research. Moreover, the results of R and D may be wasted unless accompanied by other policies in the economic and institutional sphere.

The long-term characteristics of energy R and D create problems in the area of how to decide on projects known to have payoffs only in the very long run, if at all. Applying standard time discounting techniques leads only to the rejection of long-term R and D projects of this kind. Also, some energy R and D schemes, perhaps solar energy research for instance, are primarily a safeguard against a catastrophe that is most unlikely to happen—the failure of everything else. If fossil fuels are either depleted or too costly to extract, if the technical and safety problems of nuclear power are intractable, if the scope of geothermal energy is too small, if the practical and cost aspects of hydrogen as an energy source cannot be resolved, it might be necessary to invest hugely in the ultimate source—solar power. The cost of simultaneous failure of all these alternatives would be very, very high, but the probability of occurrence is very low. Most analogous risks of this kind (household fire protection, damage to a starlet's physical assets) would be dealt with in private insurance markets. But who would insure against threats to the survival of the economy as a whole, or even civilization itself? We insure ourselves by investing in the R and D projects that protect us against the threats. But how do we decide upon the appropriate premium (the level of R and D spending in this area)? The costs of catastrophe are beyond calculation, the probability of catastrophe cannot be estimated, and—unlike almost all insurance policies—payout cannot even be guaranteed if the catastrophe occurs.

There are no hard estimates of energy R and D spending in the United States. The federal budget provides details of expenditures by agency, e.g. AEC, EPA, or by category, e.g. "energy development and conversion," "environment," but they are incomplete. Table 11-1 shows approximate energy R and D expenditures in both the private and public sectors for 1963, 1973, and 1975. The 1975 estimates are particularly flimsy, since the private sector figure is obtained by applying a growth rate to current estimates, while the public sector estimate is merely the recommended level of spending in a report prepared by the chairman of the AEC scaled down to 1973 prices. Nevertheless, the figures provide some guide as to orders of magnitude. Important features are the faster growth in public than in private spending since the early 1960s and the rapid acceleration in both since 1973. This will be accentuated by the repercussions of the decision to spend $10 billion on federal energy R and D over a five-year period, beginning in the fiscal year 1975. This expansion is accompanied by a decision to centralize the control and coordination of all energy research under a new agency—the Energy Research and Development Agency (ERDA).

The distribution by energy source varies significantly between the private and public sectors. Most of the oil and gas research has been carried out by private firms or, to a lesser extent, industry-wide institutions such as the American Gas Association which supports research through the Institute of Gas Technology.

Table 11-1
Energy R and D Expenditures, 1963, 1973, and 1975

	$m.			Average Annual Growth Rate (%)	
	1963	1973	1975[a]	1963-73	1973-75
Private	857	1000	1200	1.5	9.5
Public	460	627	1389	3.2	48.8
Total	1317	1627	2589	2.1	26.1
Of which in private sector:					
Oil and gas	469	600			
Electricity	371	300			
Other	17	100			
In public sector:					
Oil and gas	56	26			
Nuclear fission	293	356			
Nuclear fusion	36	65			
Nonnuclear electric	56	b			
Coal	15	94			
Environment	–	55			
Other	4	31			

[a] = estimated at 1973 prices
[b] = included in "other" category
Source: Based on J.E. Tilton, *U.S. Energy R and D Policy* (Washington, D.C.: Resources for the Future, 1974), pp. 1, 14, 19.

Public sector energy R and D has been dominated by nuclear power—67 percent of the 1973 total, and almost two-thirds of this were spent on the LMFBR. Nuclear energy accounts for 6 percent of total government R and D spending in the United States; a lower share than in many other developed countries, such as France (18 percent), West Germany (17 percent), the United Kingdom (12 percent), and Japan (8 percent). Research into coal technology and environmental aspects of energy consumption have recently received increasing attention from the federal government. In the private sector, a very high proportion of the research has been in development or in applied research, not basic research. The electric utilities have been poor sponsors of research historically, but the recent establishment of the Electric Power Research Institute (EPRI) is intended to boost research in this sphere. Apart from oil and gas and electric power, most private research is widely spread in such fields as coal gasification, geothermal energy, conservation, and solar energy. The federal government has neglected nonnuclear electric power, despite the scope for technological improvements in this sector.

An interesting question is whether energy R and D has been too high or too low. Since we cannot operationalize the theoretical framework, this question cannot be answered satisfactorily. With the benefit of hindsight, it may have been too low in the past. Relatively falling prices provided little stimulus to research. Also, there is scope for technical progress in the energy sector rivaled in very few other fields. On the other hand, the recent expansion of the R and D program may be too rapid, and introduced for the wrong reasons. The $10 billion decision was apparently made *prior* to any determination of national energy goals. The problems of 1973-74 were due far more to bottlenecks in plant, equipment, and trained manpower in the production and transformation of energy than to difficulties that might be solved by R and D. If there are short-term goals, insufficient consideration was given to alternative solutions other than expanding the energy R and D budget. With 54 percent of the total energy R and D spending located in the public sector, more than twice the government share in other nondefense and nonspace R and D, it can be argued that it would be more efficient in terms of type of research and availability of manpower and other resources to have a higher share in the private sector, though perhaps with government support or participation. The planned rate of increase in federal energy R and D is probably too fast for maximum effectiveness. Since there is no organized opposition group to increased spending on energy R and D, it is possible that there has been an overreaction under political, "expert," and media pressures. Promises of more research spending is an easy political decision compared to, say, compulsory restrictions on energy use.

More government spending on energy R and D probably means less spending on something else. Little examination has been given to the opportunity costs of resources applied to energy R and D. Sometimes it is argued that fiscal constraints could be avoided by having a tax on energy consumption and using the receipts to finance energy research. Superficially attractive, this suggestion has major drawbacks. Earmarked taxes reduce budget flexibility. In view of the long-term payoffs, the beneficiaries from R and D in energy are a very different set from the payers—today's energy consumers. This proposal is merely a device to take energy R and D out of its proper context, where it would be subject to general resource allocation criteria.

The consolidation of government research efforts in ERDA has good and bad points. Its main benefit is to facilitate a more comprehensive evaluation of all potential energy R and D projects than when responsibility was shared among a proliferation of agencies, and thereby possibly to improve the allocation of funds. On the other hand, there is a danger of a large agency having too much independence and power and being allowed to operate separately from the determination of energy goals and instruments as a whole within FEA. Other side effects might include reducing incentives to conclude R and D projects quickly, inhibiting the rapid transfer of new technology from the laboratory into production, discouraging private sector research, and eliminating alternative sponsors for promising but risky projects.

These arguments do not demonstrate that energy R and D is too high. However, they suggest that the decision to raise federal spending in this area was made too hastily without close examination of the goals or of the alternative means available for achieving these goals. More energy R and D may be justified as a long-term strategy for the attainment of long-term objectives, but not as a device for supplying immediate solutions to the problems, bottlenecks, and need for adjustments that suddenly occurred in 1973.

Energy Taxes and Subsidy Policies

Intervention by the government in the energy sector via taxation policy has been traditional, particularly because of preferential tax treatment for petroleum producers and, to a lesser extent, for other mineral producers. It is also a topic of perennial and recurrent interest, especially in the light of recent controversies about a tariff vs. gasoline tax and other government options. Elsewhere in this book (see pp. 143-45) there is a discussion of gasoline taxes and horsepower taxes as possible instruments for achieving automobile fuel conservation; here more general considerations are raised.

In a recent detailed study of energy tax policy, Brannon[2] argued that the optimal path was one which gave maximum opportunity to consumers and producers to respond to changes in marginal utilities and marginal costs. In other words, he argued for the restoration of the market price system in the energy sector. He argues that the energy industries are workably competitive,[3] a position that some might wish to challenge, and in any event the way to counteract monopoly is by antitrust policies, not taxation. This viewpoint, acceptable in the abstract, ignores the fact that the antitrust machine moves ever so slowly in today's conditions, while tax measures are executed relatively swiftly. Moreover, Brannon adopts the familiar posture of market economists that it is bad to keep energy prices low on the grounds that this would hurt the poor. Because the rich spend more on energy, they have more to gain from low prices. If rising prices have adverse effects on the poor, the appropriate remedy is via income transfers. Again, this position takes little account of politics. The poor not only spend a higher proportion of their incomes on energy, but also their demands are more inelastic. As for the argument that distributional objectives should be dealt with by direct income transfers, this may be impeccable economics, but in the real world the transfers never seem to take place, particularly in contemporary America. If this strategy were to be recommended, it would be more viable if the transfers were energy related, such as energy tax credits and/or energy stamps. Yet even a market-oriented study recognizes the need for some tax and subsidy interventions, particularly to deal with pollution.

Brannon's general conclusions can be summed up under three headings: elimination of distortions resulting from energy taxation; the gains from higher energy prices consequent upon tax reform; and the pros and cons of specific energy taxes. Especially before the recent changes tax provisions operated unevenly between energy sources: 13 percent of market price for oil and gas, 1 percent for coal, nothing for hydro- or solar power. These tax inequities cost the Treasury about $4 billion, and at least one-half of the benefits accrued in the form of higher royalties and profits rather than lower prices. Foreign tax credits needed tightening up, because they went far beyond relieving double taxation of producers' income. Tax benefits to publicly owned electric and gas distribution systems (the absence of income tax on the return to capital invested and access to tax-exempt bond financing) hold prices down unnecessarily. Some, but not all, of these distortions have now been eliminated.

The result of abolition of all tax concessions would be higher energy prices. This would be desirable to dampen demand as well as providing compensatory incentives to producers and royalty owners. Higher tax revenues might be used for redistributional purposes. Restoring free competition might stop short of free trade, but a tariff reinforced with an emergency stockpile would be preferable to a quota from the point of view of maintaining a degree of competition.

As for specific taxes and subsidies, Brannon rejects general excise taxes because they are indiscriminate, whereas there is a stronger case for concentrating taxation on energy uses that involve unpaid-for social costs. Thus, there might be a case for more automobile taxation, preferably via congestion highway tolls and central city parking taxes rather than by higher gasoline taxes (this suggests that automobile taxation is conceived in terms of traffic congestion costs rather than as a device to promote fuel economy). Pollution taxes would be more efficient than the present complex regulatory system to deal with high sulfur emissions and other problems. The case for a mixed strategy (taxes and controls) is strong, because of the increased flexibility (variable tax rates according to the severity of the problem, regional differentiation, uniform taxes with emergency stand-by controls, etc.). Subsidies for energy R and D may be justified, especially for research into the scope for energy savings. However, if subsidized research increases the profitability of certain fuel resources, it would be appropriate for the government to recoup windfall profits via special severance taxes.

This recipe for action relies heavily on the effectiveness of the market. Interference with market mechanisms is primarily justified in terms of external economies. The advantage of this approach is that—apart from political obstacles—it is easy to introduce and simple to operate. Its major disadvantage is that there may be interdependencies between different segments of the energy sector and between energy supplies and the welfare of the economy which are not solvable via relative price changes alone or by ad hoc manipulations of the price system via taxes and subsidies. The energy sector in current circumstances raises issues of a technical, social, and political nature as well as matters of economics. It is at least arguable that these issues can be resolved, if at all, only within the framework of a comprehensive and coordinated national energy policy.

The Federal Power Commission

A major characteristic of the energy sector is that it is not only an interdependent mix of quite distinct industries, but is subject in several important respects to regulation by the federal and, to a lesser extent, state governments. Of the regulatory agencies the most important is the Federal Power Commission, established in 1920. Under the Federal Power Act of 1935 the Commission was transformed into a regulatory agency with the responsibility of assuring a plentiful supply of electric energy at "just and reasonable" prices, and control was extended to natural gas producers in 1938 (the Natural Gas Act). Much later, as a result of the 1954 Supreme Court decision in *Phillips Petroleum Co. v. Wisconsin*, the Commission began to control the prices charged by natural gas producers on sales to interstate pipelines (see also pp. 83-85). The goals of the FPC have related to attempts to hold down prices to consumers (a distributional goal) and to maintain and promote coverage and quality of service (an efficiency goal). In some cases these goals are compatible. For instance, antimonopoly actions can lead to lower prices and increased output. However, if control keeps price levels below those that would prevail in a free market the consequence will be shortages due to excess demand and reduced supplies. The methods used by the FPC to control prices have been time-consuming and cumbersome, and its administrative costs have been heavy. Most aspects of the rate-setting process have been dealt with via an adjudicatory process which works best if the choice is to be made between two or a limited range of outcomes, whereas many rate-making problems raise a large number of options. Settlement of disputes is a more suitable goal to be handled by adjudication than the aim of efficient resource allocation.

The FPC has been widely criticized for its failure to allow natural gas prices to rise in the face of growing shortages, and for acting belatedly and arguably too conservatively only in 1974. In a recent study Breyer and MacAvoy have indicted the FPC on a much broader front:

In the late 1960s, with the commission operating at full steam, results were dismal: prices collected by the pipeline companies were not perceptibly lower than they would have been without regulation; setting field prices for natural gas did the residential consumer more harm than good by affecting the market so as inadvertently to bring on a gas shortage; and with federally regulated sales constituting only a minor portion of electricity sales, manifold opportunities to shift costs tended to render federal pricing ineffective. Commission planning efforts faltered.[4]

Their analysis is, on the whole, convincing.

Policy Proposals of Three Major Studies

Although the policy conclusions of three major studies (Project Independence, Ford Foundation, MIT Energy Laboratory) overlap to a considerable extent, they differ in emphasis and, indeed, in philosophy. The MIT study[5] strongly favors market prescriptions, price incentives and fiscal adjustments. The Ford Foundation approach[6] is a bombardment of supply, demand, and conservation measures, indiscriminate in the sense of paying little attention to priority or political feasibility. In general terms, many of their prescriptions are regulatory and strongly interventionist. The FEA's Project Independence Blueprint[7] is much more cautious. In part, this reflects the natural reticence and relative blandness of government reports; in part, it avoided committing the administration before detailed strategies had been worked out. It is interesting that none of these studies would have supported the energy "plan" proposed in the President's State of the Union message of January 1975—high tariffs on imported oil to push domestic prices above the world level and postponement of target dates for achieving environmental standards, particularly on vehicle emissions.

PIB offered a range of policy options, without outlining a preferred course of action other than suggesting that the most desirable strategy might be a mix of the alternative policies. There were, in PIB's view, four main options: increase production from domestic sources, taking appropriate action to accelerate supply; restraint of domestic energy demand; the development of domestic plans to avoid disruption resulting from import supply cuts or sudden price increases; international action to reduce the likelihood of an interruption to imports. Choosing from among the various options is made more difficult by the uncertainties of the situation, especially the degree to which price rises may dampen demand, the ultimate production potential of new oil fields, and the lead-times for delivery of new energy production.

Despite its decision to avoid a commitment, PIB commented on some of the policy choices and on the influences upon them. Higher world oil prices make attainment of self-sufficiency easier, but the United States would nevertheless be much better off with lower prices. Many policies to reduce price uncertainty such as tariffs and subsidies are cumbersome and inefficient. Any chosen energy policy must be flexible enough to adjust to the dynamics of the situation. Domestic policies might have repercussions on world oil prices which could abort the original policies. For instance, accelerated supply strategies at high world oil prices would reduce imports, putting downward pressure on world prices that could make the new domestic investments uneconomic. Nevertheless, stimulating domestic supply would be practicable provided that world prices did not fall below $7, though hard environmental trade-offs could not be avoided. Restraint of domestic energy demand is appealing because of its conservation benefits and its feasibility without sacrificing economic growth. Its drawbacks, in the eyes of PIB, are more government intervention and restrictions on individual choices and life styles. Reducing imports would be difficult and, beyond a certain level, costly. Zero imports are not a desirable policy goal since there are cheaper and more efficient means to reduce vulnerability to import disruptions.

PIB believes that oil-exporting countries will be faced with an increasingly stagnant international market if current price levels continue into the 1980s. To meet this situation and to protect against the threat of embargoes, there are several possibilities. One is by cooperation among oil-consuming countries, e.g. the International Energy Program (IEP), to combat the OPEC cartel and to agree on sharing available supplies in cases of supply restriction. Much more feasible is a domestic emergency stockpile program in which stockpile levels could be adjusted sequentially according to what happens in the world oil market (see pp. 132-36). The advantages of a storage strategy seem so overwhelming that the disadvantages should not be forgotten:

It may take several years to design and implement a major storage program, yet our vulnerability is highest now and storage may be of little value five to ten years from now. Secondly, building storage will require greater imports now. Purchases in the world market for storage will tend to sustain higher prices in the short run and put additional strains on the international financial system. Finally, if we purchase storage now to avoid the risk of a large economic loss from an embargo, we also risk a possibly large capital loss if the oil we store drops dramatically in price.[8]

PIB admits that there is little scope for international supply diversification by 1985. Apart from storage and international cooperation, emergency demand restrictions and supply allocation can help substantially to reduce the adverse economic impact of an embargo.

The MIT study identifies two separate strands in pursuit of a self-sufficiency strategy: reducing the risk and costs of import disruption and its impact on foreign policy and the domestic economy; and relieving the burden of high-priced oil imports, if the world price is higher than the long-run cost of increasing domestic supplies. Any strategy runs into problems. For instance, the normal method of dealing with shortages is to allow prices to rise so as to choke off demand and stimulate supply. In the energy sector, however, its pervasiveness and the magnitude of the price increases required would involve massive income transfers from consumers to owners of energy resources. The simple strategy of promoting the development of coal would conflict with environmental goals, while converting coal to synthetic fuels substitutable for oil would be intolerably expensive in the present state of technological knowledge. Attempts to achieve independence would be more costly the faster they are put into effect.

In these circumstances, the MIT study proposed policies under four headings: changes in federal regulatory policies; actions to improve the effectiveness of market mechanisms; income distribution measures; and security against import disruption. The first line of action included: deregulation of the field price for natural gas; Interstate Commerce Commission (ICC) action in railroad rate cases to inhibit transport cost restrictions on western coal deliveries to the East; more effective leasing of federal lands containing energy resources; possible revisions in the sulfur pollution standards of the Clean Air Act; and speeding up licensing agreements for nuclear power plants.

Suggestions on the pricing front rejected tariffs and quotas to boost prices artificially. Unless world prices dropped dramatically, current prices would offer more than enough incentives to increase domestic supply. The exception is the synthetics field, where incentives might be provided by federal purchases at a guaranteed price. In view of the wide gap between prices on "new" and "old" domestic oil, a phased relaxation of price regulations on old oil was recommended (this has now been endorsed by the administration), provided that action was taken to eliminate windfall gains to the oil industry. The MIT study also gave some support to measures to promote efficiency in energy use, such as labeling of equipment, action to stimulate capital substitution for energy (e.g. insulation of buildings), controls on the use and availability of energy-use-intensive devices, and selective excise taxes to promote energy conservation.

Income inequities resulting from more reliance upon market forces could be dealt with by variable excise taxes on domestic crude, introduction of an excess profits tax, and bringing the oil companies back into the mainstream of the corporate tax system (including the abolition of oil depletion and other tax concessions). The protection against import embargoes could be obtained via a stockpile policy.

The Ford Foundation policy recommendations were based on the desire to achieve its Technical Fix scenario, involving cutting the rate of growth of energy consumption to below 2 percent. This implied an energy conservation strategy, especially via the use of technology to save fuel. This could be achieved at little cost to economic growth, would save $300 billion of capital expenditure over the next twenty-five years, and might permit a switch to ZEG (Zero Energy Growth) after 1985. The three main economies would be: energy-saving techniques in the construction and operation of buildings; better gas mileage for cars; and greater efficiency in industrial plants via investment in new technology and reducing heat waste. The study argued that specific actions on a wide front were needed to achieve the TF energy growth path.

Particular importance was attached to energy conservation measures. These included elimination of promotional discounts and full-cost pricing of electricity, national policies to promote the manufacture and sale of more efficient automobiles, provision of incentives to increase energy efficiency in space heating and air-conditioning of buildings, and initiating government programs to promote technological innovation in energy conservation. An institutional suggestion was the recommendation of establishing an Energy Policy Council within the Executive Office of the President to develop and coordinate national energy policy.

These measures would be accompanied by action aimed at increasing energy supply by 28 percent, 1973-85. Before 1985 it would be sufficient to rely on existing sources and technology: new onshore and offshore oil discoveries, especially in Alaska and the Gulf of Mexico; raising coal output from deep mines and via strip mining where reclamation is possible; the entry of power plants already planned or under construction; and secondary and tertiary recovery of oil and gas from existing wells. After 1985 it would be necessary to resort to less conventional sources, perhaps at least two from oil imports, nuclear power, Rocky Mountain coal and shale, and more drilling in the Gulf of Alaska and offshore around the East and West coasts. In addition, coal output would have to double by the year 2000. It would not be necessary, on the other hand, to continue the expansion of the electric power industry already planned, since electricity consumption under TF would grow at a rate only one-half of that experienced historically. The Ford Foundation believed that prices were more than adequate to provide the required incentives, but there was a need for eliminating special tax advantages, coordinating oil and natural gas pricing policies, establishing a 90-day-supply oil stockpile, and redirecting energy R and D into nearer-term opportunities such as geothermal energy and into environmental problems.

Social equity problems might be dealt with by explicit action to help the poor such as energy stamps, special grants, or fuel allocations. However, it was important to save *all* consumers from excessive energy costs, while the difficulties of the low-income groups were perceived as being much wider than the effects of higher energy prices requiring national redistributive measures on a wide front. Employment and economic growth need not be adversely affected by energy conservation provided that investment tax incentives were devised in a manner that promotes energy use efficiency. Also, policies relating to housing, transportation, R and D, and the environment could, and should, be consistent with conservation objectives. Although the study recognized that energy-environmental trade-offs depended on value judgments, its view was that air pollution standards should not be sacrificed, nuclear power should be soft-pedaled until the safety problems are resolved, and that attention should be given to the environmental repercussions of using western coal and shale and extending offshore oil development. It was opposed to the LMFBR program, and believed that there had been underinvestment in research into renewable resources such as solar energy. In the utilities sector, pricing policies should be reformed, regional should be substituted for state utility commissions, and electricity generation and transmission ought to be separated from local distribution. Resource assessment, planning, and regulation required reform, particularly the introduction of a competitive leasing system for federal lands and resources. The political dimension to energy policy was recognized by arguing that campaign finance reform was important to weaken the oil industry's political strength, while the role of citizen participation in influencing energy policy could be critical.

Finally, the study stressed the international aspects of energy policy. The dramatic rise in oil prices and the associated shift in power relationships had destroyed the case for leaving trade in oil to the market. Although the future course of the world oil price was difficult to predict, a sharp decline was most unlikely in view of the strength of the OPEC cartel and the short-run lack of substitutability of other fuels for oil. In this climate agreements should be sought to stabilize the impact of oil trade on the world economy and to minimize the unfavorable impacts of oil on trade and finance. Nevertheless, except in the rejected High Import option of the Historical Growth scenario, there was no reason why foreign policy problems should be considered insuperable. The study believed that energy conservation strategies were the best approach to dealing with international energy problems. Also, there should be restrictions on the export of nuclear fuels and equipment, though in the interests of peace rather than because of energy considerations.

An Energy Standard of Value?

In recent years there has been increasing dissatisfaction with using GNP in money terms as a yardstick for measuring changes in welfare. Regional, environmental, and distributional impacts of economic change cannot be evaluated satisfactorily, if at all, within the framework of such a calculus. The result has been inconclusive discussion and research about alternatives, such as social indicators. Sometimes, it has been suggested that criteria for resource allocation other than prices should be developed. One proposal is that dollar evaluations should be reinforced by energy consumption criteria. The grounds are the assumptions that energy is the critical resource in mature industrial economies and that its supply is being used up too fast. This implies what might be described as a Btu theory of value.

There is no doubt, of course, that an activity's share of energy consumption may differ widely from its share of GNP. Also, it may be arguable that some activities may use energy extravagantly in supplying "needs" that are difficult to justify on social grounds. If energy requirements were taken into account in sectoral allocation and public-policy decisions, it is possible that substantial economies in energy consumption could be achieved relatively painlessly and without distorting the structure of the economy. In production, for example, public policies might be used to promote resource transfers between energy-intensive and labor-intensive industries and technologies, particularly in cases of close substitutability. Such a strategy might be quite consistent with pursuit of employment and energy goals. To take one instance within the capital-intensive sectors, one study[9] showed (using 1963 input-output data) that a one-billion-dollar increase in primary aluminum sales set against an identical fall in steel deliveries would have increased energy use by 116 trillion Btu (0.2 percent of the total) and reduced employment by 15,000 jobs (0.03 percent of the total).

There is more scope for energy-saving substitutions within the broad area of product use. Obvious examples include intercity transportation, food, and packaging. For instance, a complete switch to returnable beverage containers would save 0.5 percent of national energy consumption, as well as increasing employment (130,000 jobs) and reducing consumers' expenditure. The automobile and related industries (again based on 1963 data) consumed 12.4 percent of GNP, provided jobs for 12.0 percent of the labor force, but consumed 20.7 percent of total energy, and in view of deteriorating fuel economy these differentials have probably widened since 1963. As for general consumption patterns, food, housing, and transport account for three-quarters of household direct and indirect energy use but only 47-58 percent of household income (according to level of income). The relationship between changes in life styles and energy consumption is a question that deserves further research, the results of which might be very useful in assessing the feasibility of household conservation programs.

The scope for governmental control of energy use is considerable, even though many of the instruments may be politically unacceptable. These controls could take any of the following forms: pricing and taxation, rationing and other mandatory allocation devices; information and educational strategies; public-investment decisions; and land-use controls. Given the potential for fuel economies and a mix of strategies to promote them, does this provide a case for enthroning the energy standard of value?

The sensible answer is a negative one. It is hard to establish that energy resources are so critical that they should become the dominant criterion for policy and expenditure decisions. If the justification for this is based on supply constraints, these constraints should be reflected, more or less, in relative price changes which should guide resources in the appropriate direction. Despite the defects of the monetary index, it remains the only common standard available. It is true that in a world of governments, monopolies, spatial frictions, and inertia, second-best problems are overwhelming. But to substitute Btu-minimizing criteria would misallocate resources even more severely. More modest proposals, such as the preparation of a national energy budget, are more attractive if they are regarded as an information base to improve policy decisions rather than as an instrument for arrogating a single policy goal to a position of unrivaled supremacy. Such a step would be to give way to special pleading rather than to lead to rational policy formulation.

12 Toward a National Energy Policy

Diagnosis

Any assessment of what the United States should do, or not do, by way of energy policy must depend on how the situation is diagnosed. Unfortunately, there is no general agreement on this, mainly because there are so many uncertainties and imponderables, but also because national goals have been loosely specified. In any event, there is no reason why the analyst should accept the officially stated goals without question. Assuming that goals are multidimensional, trade-offs arise that can be resolved—in the absence of a measurable social welfare function—only by the exercise of value judgments. There is not even clear-cut agreement about how the main parameters are interrelated. For instance, what world oil price (assuming that this is the price setter for energy industries at the present time) is necessary to provide sufficient incentives to exploration for fossil fuels and to R and D investment in new technologies and alternative fuels? The answers to this question differ among industry spokesmen, members of the administration, scientific experts, and economists. The floor price to stimulate exploration in oil and gas is much lower than the floor price to justify investment in oil shale or synthetics, so much depends on forecasts of the fuel mix in, say, ten years' time.

Even if a stated objective is accepted at its face value, it may have been proposed for several different reasons, and the relative weight of these may indicate that some instruments are more suitable for achieving that objective than others. For example, the aim of energy independence has been justified in several ways. Two of the most important have been the security argument—whether couched in terms of the need for an independent foreign policy or because of fears of damage of supply interruption to the domestic economy—and the balance-of-payments costs. If either of these were dominant, it would suggest quite a different set of measures. In the first case, a large stockpile and free imports might be an efficient solution, but it would have unfavorable implications for the balance of payments, especially because building up the stockpile would require much higher imports in the short run. If balance-of-payments criteria were crucial, the preferred strategy would require a drastic cut in the import bill—either by a physical quota or, provided that there was sufficient elasticity in the demand for petroleum, via a tariff. Also, an instrument may serve more than one goal in practice, if not in the theory of economic policy. The higher domestic prices resulting from the imposition of tariffs or quotas may be regarded not as an undesirable social and economic cost but as an effective weapon for dampening demand and promoting conservation of fuel. This advantage may be especially appealing to an administration with ideological qualms about raising the tax on gasoline, the most obvious alternative for inducing conservation.

Another reason inhibiting diagnosis is the uncertainties of the future, particularly the durability and the effectiveness of the OPEC cartel. The analysis in Chapter 8 has suggested that the cartel is unlikely to break up in the near future, partly because of its past success, partly because the ability of the producers to cut back supply is much greater than the ability of the consumers to reduce their demand, partly because concerted action by the oil-consuming countries could develop and would legitimize the existence and operations of the cartel. If this assessment is wrong, and it is a matter of speculation or, at best, judgment, the prescriptions would be different. Despite pessimism about the collapse of OPEC, this study does not argue for a draconian approach to energy policy. The costs of overreaction—misallocation of resources, inequity among consumers and between consumers and producers—are far higher than the costs of doing little or nothing. If the cartel were to break apart, the case for a national policy based on higher prices and import restrictions would fall to the ground, except that the protection of an emergency stockpile might be a judicious insurance against a renewal of the conditions of late 1973.

The implication of the arguments outlined in this book is that there is no lasting energy "crisis" in the United States. The serious problems apply only to oil. Even the notion of an oil "crisis" is unconvincing, since there is no oil shortage, either in the short or medium runs. A more accurate interpretation is that the world in general, and the United States in particular, naturally found it hard to adjust to a very sudden change in relative prices, between oil and other fuels and between fuels as a whole and other goods and services. This readjustment was more difficult for some countries including the United States because the price changes were accompanied by politically motivated restrictions in supplies. Readjustment was aggravated by the lack of substitutability among fuels in certain energy end uses, notably transportation. However, the origins of the changing situation had little to do with the risks of energy resource depletion, but were the unexpected and accidental product of a loosely connected chain of events from Tehran in 1971 to the Arab-Israeli war of 1973. Accordingly, to react to the situation as if there were a permanent and full-scale energy crisis of the resource extinction type is inappropriate, unless one subscribes to the view that these events brought home to the world the underlying realities. The latter position is tenable only if one believes that the scope for further oil and gas exploration is severely limited, that coal is unusable on environmental grounds, and that the technological possibilities of developing nonfossil fuels have been grossly exaggerated. If there are grounds for optimism on any one of these three counts, many of the prescriptions suggested in the

United States in the past two years are unsound because they assume the existence of a serious, long-term crisis and fail to zero in on the real problem—the dependence of an energy-intensive and automobile-reliant economy on a fuel that suddenly became relatively expensive after decades of being very cheap.

This position does not imply that there have been no major, and perhaps lasting, changes in the world and the local energy situation. There has been a major shift in world economic power. The years of ultra-cheap energy are probably over. Life styles may have to change. The required readjustments may call for political, social, and economic rather than merely technological solutions. The repercussions on the international balance of payments, though less intractable than expected, have still not worked themselves out, especially since the appetite for arms and technology of the oil producers must be finite.

Similarly, this analysis is not a recipe for inaction. The events of 1973-74 did expose a degree of vulnerability in the American economy and way of life, and there are some lessons for policy-makers. For example, it was realized that the very high per capita consumption levels of the United States compared with the rest of the world have costs in addition to the gross inequities. Wasting resources is inefficient even for the affluent. Thus, the case for an energy conservation program and for adapting life styles in a way that is not too damaging to welfare and utility levels is strong on general economic and social grounds, independent of any arguments about energy supply inelasticities. Moreover, the experience raised certain issues of energy policy that are now discussed in the public forum rather than merely in the journals of experts or the committee rooms of Congress. These include: the appropriate level of tax incentives for oil companies and other energy producers; the effectiveness of the regulatory operations of the Federal Power Commission and other agencies; the scale, scope, and direction of energy R and D spending; the income and welfare distribution aspects of fuel taxation; the trade-offs between energy use and environmental protection, and its implications for national environmental strategies; the potential conflict between individual liberty (as reflected in spending behavior and life styles) and public interest; the international obligations of the United States to other countries, especially industrial countries without domestic oil supplies; the ethical and social problems posed by a rapid extension of nuclear power; and many more. Since solutions to so many of these problems can be implemented only by government initiative and policies, the energy debate revealed new contradictions between the market-oriented ideology of American society and the responsibilities of government in the modern world.

Furthermore, the trauma of those months may have had salutary effects on the American people. The lines at the gas stations and the intermittently empty pumps gave the American consumer a small pang of the privations that afflict, more frequently and more seriously, so many in the rest of the world. The experience of that great leviathan, the American economy, grinding slowly at the behest of a clique of distant, anachronistic desert sheiks was a far more sobering lesson in international interdependence than a million political speeches. The current pressure for energy independence is a blind, if understandable, reaction to that experience. Unfortunately, the lesson (perhaps with another example) may have to be repeated before its real point sinks home.

Energy Independence

The nebulous concept of energy independence is the principle by which each of the strands in national energy policy is judged. What energy independence means precisely is unclear, since the statements of the administration have varied so widely, and why it is important is also subject to many different interpretations. Sometimes the argument is that the United States should aim for zero imports by 1985. As shown by PIB, this would require the combination of a high world oil price and an accelerated supply strategy, perhaps reinforced by demand conservation measures.[1] If the world price is below the domestic price, the costs of a zero import policy may be very heavy. On the other hand, if the world price is higher than the domestic price, the justification for protection for domestic producers is unclear. For instance, most observers believe that the current gap between US production costs and the world price is more than sufficient to stimulate increases in domestic production. One argument, frequently made, is that unless domestic producers are protected their investments in new production facilities can be wasted by undercutting by the low-cost, though high-price, foreign supplier. However, this does not provide a case for permanent protection, but merely for enabling powers to impose a tariff or other protective device if the undercutting took place.

There is no evidence that the energy independence objective is based on the high price of imports. If that were the case, it appears odd that the problem should be fought by making prices even higher. If higher prices are judged desirable to dampen demand and to promote conservation, this has nothing to do with an import protection policy. The administration's preference for a tariff rather than a higher gasoline tax shows the irrelevance of the relative import-domestic price argument. An equivalence could be made between a tariff and a tax increase from the point of view of raising domestic prices so as to depress demand. The major difference between the two approaches is that the tax measure does not alter the relationship between the domestic and the foreign supply prices, and the price increments accrue solely as increased government revenue. A tariff, on the other hand, alters the relative price ratios, and makes investment in domestic production profitable even when the output could only be sold at a price higher than the pretariff foreign supply price. Also, part of the price increment accrues to domestic producers unless it is taxed away by an excess, or "windfall" profits tax. More serious is the fact that a tariff results in resource misallocation by attracting too many resources into domestic energy production.

A modified variant of the energy independence objective is a phased reduction in imports stopping short of zero imports and with an imprecise target reduction. For example, the administration's proposals of early 1975 argued for a 1-MBD cut in oil imports in 1975, a reduction of 2 MBD by 1977, and a reduction in imports to about 20 percent of demand by 1985 (estimated to be 4.7 MBD). The House Ways and Means Committee's proposals of the same period called for a more modest short-run cut (by 0.5 MBD in 1976) but a similar longer-term reduction (to 25 percent of demand by 1980 or soon after). These are much more feasible and sensible targets.

The disagreement between the administration and Congress about energy policy in the spring of 1975 reflected differences about choice of instruments rather than about general objectives. Both agreed on the following necessary elements in an energy policy: a national effort to conserve fossil fuels; a limit on oil imports; higher prices for petroleum products; and government stimuli to the development of alternative energy sources. The appeal of a pricing strategy was that it could be used to support all of these principles. However, the simplicity of such an approach is no guarantee that it is the most efficient solution. The administration's preferred instrument was a three-phase tariff of $3 per barrel, supported by supplementary measures such as gradual relaxation of price regulations on domestic fuels and postponement of automobile emission standards in return for attempts to achieve greater fuel savings by 1980. One House proposal (the Wright-Pastore scheme) called for a 5-cents-per-gallon tax increase, a cut of 0.5 MBD in oil imports, and a graduated excise tax and rebate scheme imposed on cars according to fuel economy. The gasoline tax revenues would be used to finance energy R and D, incentives would be provided for private home insulation, and a National Energy Production Board would be established to direct US energy development. The Ways and Means Committee believed that these proposals were too weak, and developed its own, somewhat tougher, program. The key elements were a steep gasoline tax and a more sophisticated import quota scheme. The gasoline tax would be raised in steps of 10 cents per gallon to reach 40 cents by 1979, but the regressive impacts would be minimized by imposing the tax only on purchases in excess of a basic allowance of 9 gallons per vehicle per week. This measure would be reinforced by a tax-rebate scheme to promote purchases of fuel-efficient cars. The quota scheme would be a "sealed bid" scheme under which the government would periodically auction off a limited number of import licenses to the highest bidders, administered by a federal purchasing agency. The idea would be to encourage individual OPEC producers to cheat on the cartel by deviating from the established OPEC price. The differences between the proposals reflect ideological attitudes of the political parties more than a dissimilarity of approach: the emphasis on price and market incentives, including higher profits for producers, in the administration's proposals contrasts with the taxation strategy of the Democrats with greater stress on equity considerations. Similarly, the tariff is a pricing instrument, whereas the quota scheme and the National Energy Production Board imply an extension in the functions of government.

It is unclear why energy independence per se should be considered such a critical goal. To escape from the domination of high import prices would be understandable, but the fact that all the measures suggested to promote independence include action to raise prices still further shows that the price argument is not the key factor. A variant of the import cost hypothesis is the balance-of-payments cost. The US balance-of-payments deficit reached $10.58 billion in 1974. Although there were several contributory factors, easily the most important was the increase in the oil import bill by $18 billion, which was only partially offset by expanding exports. However, in comparison with the balance-of-payments experiences of other industrial countries, the US deficit was bearable relative to the size of GNP. This raises a much more general point. US dependence on oil imports is far lower than that of most other industrial countries in the sense of imports' share in total demand. A central feature of the modern international economy is the high degree of economic interdependence among countries. To retreat into autarky is no answer to the problem, however strong the desire to avoid cartel prices.

Many of the arguments used to support the official objectives stress security considerations. However, it is important to appreciate that these do not require independence from imports at all, merely some method—of which the emergency stockpile (pp. 132-36) is the simplest—of insuring against supply interruptions. Moreover, although the damage to the economy risk is the most serious aspect of security, there is little doubt that for the Administration security means freedom to pursue an independent foreign policy—a view that has been strongly influenced, of course, by the experience of the 1973 embargo. However, the argument is of dubious value. The influence of the United States in world affairs is not a function of its invulnerability but perhaps of its ability to act in support of moderation and reconciliation. Many people outside America, and an increasing number within it, believe that benefits would result if the United States were not free to assume a major role in foreign policy. In any event, the constraints on her freedom of maneuver are much greater than, say, in the 1960s, when political clout was reinforced by the supremacy of the dollar.

It is widely believed that the security case for energy independence is strongly supported by the State Department, and certainly that Department has been the front runner in pressing for cooperation among the oil-consuming countries and for the "floor price" plan where consumers would guarantee a floor price to protect new domestic investments in energy supply from undercutting by OPEC or other foreign suppliers. The problem with this strategy is that it makes much more sense for the United States than for the other industrial countries. The United States has many marginal producers to protect, and the governments of most European countries have a much more active role in the energy sector, and hence do not depend on the market incentives that are so critical in the United States.

There are grounds for the belief that the pressure for energy independence in the United States has much deeper roots than those mentioned by policymakers. In particular, there is a psychological dimension to the problem. American society has been built upon the concepts of independence and freedom from external pressures. The involvement of Americans with the rest of the world has always been conditional with the right to withdraw rather than on a mutual basis. The embargo and higher oil prices of 1973-74 were traumatic because they revealed that, for the moment at least, this independence and self-reliance was a figment. The idea of the United States being at the mercy, in a very limited sense, of a few small Arab nations is anathema to Americans as a whole, not merely American Jews. Much of the underlying support for energy independence derives from the ill ease of Americans at the thought of their vulnerability being exposed once again. Of course, there is no rational justification for this feeling, since all countries are dependent on others for key supplies, markets, or political support. Since supply-interruption risks can be dealt with in other ways, a policy of energy independence is based not on economic costs or even security, but rather on some "macho" desire for total independence. The justification for this policy, needed on psychological grounds, is rationalized by exaggerating the severity of the energy "crisis."

Proposals

It is obligatory for any analyst of contemporary energy problems to conclude by setting out his own recipe for energy policies. It would be priggish to back down from this challenge. Nevertheless, the following suggestions are made with many reservations. First, at the time of writing there appears to be some sense in a policy of "no action." There is a serious danger of overreacting by committing investments to unnecessary and unproductive uses. Plans to coordinate the actions of consuming countries might result in stability of the wrong kind, namely the permanency of current OPEC prices. Action to increase gasoline and other fuel prices in the short run has regressive impacts on consumers, allows producers to reap windfall profits without any guarantee of much additional stimulation to new discoveries, and is unlikely to provide a strong enough incentive to exploration. On the other hand, there is much in favor of coordinating strategies and programs between different fuels. Many of the more serious troubles of recent years stem from the lack of alignment between prices for natural gas and for other fuels.

Second, policy proposals must take account of political facts and social acceptability. It would be easy to recommend nationalization of the major oil companies, and such a step might have several beneficial results, but it would receive little political support and would be out of step with US public opinion. Many studies have argued for restrictions on the use of private automobiles and for massive programs for developing public transit systems. There is little evidence that this strategy is viable politically or that the provision of reliable public transportation has any chance in the current sociocultural climate of persuading Americans to part with their automobiles as their travel-to-work mode. Of course, there may be scope for research into long-term strategies for breaking down this pattern involving attention to social and institutional forces as well as to technology. Such research would need to focus heavily on the transitional policies required to achieve the given objective. To jump into programs that deviate radically from the past without adequate preparation would be to invite failure.

Third, policies should be designed so that they can be implemented incrementally, and should be reversible. The era of costly energy is so new that there is considerable ignorance of how the economy and the behavior of businesses and households react to it. Experimental programs such as demonstration tests for energy stamps recently tried in three areas in the United States are a good example of treading carefully. Pricing strategies such as peak-load pricing schemes for electricity consumption or direct road pricing for automobile use are typical examples of the type of policy instrument where demonstration programs are beneficial. A much trickier question is how to deal with energy R and D. Here a precarious balance must be struck between adequate scale and continuity of funding and the need to avoid the soaring and irreversible commitments characteristic of many long-term technological solutions; the experience with the breeder reactor is a good illustration of the R and D monster that devours its rivals. It is difficult to challenge the diagnosis that the distribution of energy R and D in the United States has been distorted by its heavy concentration in a few limited fields of development. Of course, proliferation is to be avoided, but all the nonfossil fuel technologies—apart from nuclear power—have received only minimal support.

Fourth, so many energy proposals imply an abandonment of the recent, and as yet half-hearted, commitment to environmental preservation. This commitment should be weakened only as a last resort; the struggle to obtain it was too hard fought for it to be abandoned lightly. Some official statements have given the impression that the administration is glad of the excuse to be able to backpedal on the environmental issue. It can be argued that today's energy problems—though complex and real—are not serious enough to justify this. The problem is that it is not merely a question of temporary postponement, but of loss of momentum. The resistance of business, especially big business such as the automobile, steel, and oil industries, has been so strong that it would take a massive effort even to lead these horses back to the trough, not to mention trying to make them drink. The price of energy independence is too high under any set of policy arrangements; it would be a disastrous exchange if bought with a relaxation of sulfur oxide standards, a pause in the struggle for effective automobile emission controls, the destruction of scenic countryside, and the profligate use and contamination of scarce western water.

These qualifications, coupled with the belief that energy problems, though serious, fall short of justifying the term "crisis," imply a cautious approach to the design of a national energy policy. In particular, apart from the importance of reversibility and gradualism, attention should be focused on desirable long-run "end-states" rather than on short-term responses to an "emergency." The reluctance of government to intervene in the United States is so strong that emergency pressures are frequently needed in order to overcome it. This has the unfortunate result that hurried policies are introduced that are likely to fail, so that the problem is rarely solved. However, by the time the failure is evident the short-run pressures have been relieved by the spontaneous dynamic of events. This relief is no indication that the underlying problems have been touched. A striking example in the 1960s is the "urban crisis." The race riots of that decade led to considerable public concern with the problems of the big cities and to a number of programs and policies that had minimal success. By the time failure was evident, however, the urban tensions had relaxed considerably. No one talks about the "urban crisis" today, but the social and economic problems that created it have not been ameliorated to any significant extent. The propensity of US policy-makers to respond only to "fad" issues makes one wonder whether the "energy crisis" is not the analogue of the early 1970s to the "urban crisis" of the 1960s. The danger is that policies will be designed to meet the short-run problem of the OPEC-induced increase in world oil prices rather than the more critical question of the long-term balance between the demand for and supply of fuels and its implications for the fuel mix.

The impression left by recent public discussions is that an oil import policy is the cornerstone of a national energy strategy. This argument is not convincing at a time when the world oil price exceeds the domestic price. In these conditions domestic supplies will be substituted for imports if they are available. If they are not available, no type of import controls can produce them out of thin air in the short run. Higher import prices are not needed to stimulate more domestic production. The present constraints on domestic supply are not related to a reluctance to invest at the prevailing price, but to inelasticities in the supply of equipment and skilled manpower and to lead-times. It may be desirable to reduce imports, but to use a higher-price approach to achieve this ignores the limited ability of many people to adjust their behavior and spending patterns in the short run. A sounder strategy is to promote energy conservation in general, and reduced demand for oil in particular, by a mix of mutually reinforcing measures aimed at changing long-run behavior but giving people the time and the opportunity to readjust. An import price-raising policy is a crude and inequitable, or ineffective, method for promoting energy conservation. If it is desired to reduce imports in the short run, the most direct way is a quantitative restriction approach combined with an allocation program to share available supplies equitably. It is arguable, however, that the social costs of supply restriction and its impact on the economy outweigh the balance-of-payments costs of allowing sufficient imports to balance supply and demand at current price levels. If OPEC continues to operate effectively, reducing imports does make economic sense, but the sounder approach is one that aims for a long-term rather than an immediate response.

One argument used in favor of import controls is that this is the only guarantee to domestic producers of the profitability of long-term investments, because it prevents low-cost foreign suppliers from undercutting. The weakness of this argument is that protection against "dumping" (an inaccurate term since in this case the "dumping price" will still assure producers of a handsome profit) does not require permanent protection, but can best be handled by a reserve power to impose import controls if and when circumstances warrant such action. For similar reasons the "floor price" strategy is irrelevant. If domestic energy supplies cannot satisfy demand in the foreseeable future, unrestricted imports could be allowed in to meet residual demand until producers abroad attempt to reduce their prices below those ruling in the United States market. At that time, but not before, an appraisal can be made of the trade-off between guarantees to domestic producers and the benefits of lower prices.

The risk of supply interruptions could be met by an emergency stockpile program. At the time of writing excess supply conditions make it an appropriate moment to begin building such a stockpile. Current stocks in the spring of 1975 were more than 10 percent higher than in 1974, and at their highest level since the spring of 1971. The costs of storage are quite moderate (see pp. 133-36). The costs of acquisition are high, however, at current prices, and the stockpile should be built up very gradually. The risks of capital loss are real but tolerable, given the security gains from the program. If demand conservation measures are promoted with vigor, by the time the stockpile is completed the economy would have a greater capacity to readjust to supply curtailment than in 1973-74. This suggests that the stockpile need not be very large, perhaps equivalent to three months' supply of imports from "insecure" sources. In case of a longer-term supply interruption, an emergency program could be prepared in advance to accelerate extraction from reserve capacity.

The most satisfactory long-term strategy for dealing with the import dependence problem is action to slow down the growth of demand by demand conservation measures. This is not an argument for a "hair-shirt" policy. On the contrary, the strength of the case is that energy consumption in the past has been so wasteful that, allowing for long adjustment lags, it is possible to achieve substantial savings without imposing heavy monetary costs on consumers, especially the poor, and without demanding drastic changes in life styles. The strategy would require a battery of complementary measures including education and information approaches, pricing and tax instruments, and the promotion of energy-saving research and technology. The composition of appropriate measures cannot be decided outside the context of the political process and the institutional environment, so that spelling out the details of such a strategy in the vacuum of this book would not be very productive. Some of the possibilities include: higher gasoline taxes, offset by rebates or credits to minimize the impact on poor motorists dependent upon the automobile for travel to work, shopping, and other essential trips; the proposed differential excise tax on automobiles, perhaps graduated in a way that would not only switch demand from full-size to compact cars, but also induce Detroit to produce automobiles of the Honda, VW Rabbit type (this would require bringing more direct pressure to bear on the auto manufacturers themselves); reform of electric utility pricing, using peak-load or similar full-cost criteria; abolition of discounts on bulk sales of fuels to large consumers; incentives to home and office insulation; revision of building codes and standards; compulsory inclusion of thermostats and time clocks on new space-heating equipment; subsidies to research and development into energy-saving appliances; and public information and education campaigns. It may be difficult to obtain agreement on a wide enough range of instruments, and they would not have immediate effects. Nevertheless, most of the proposals are relatively simple in design, and are not costly once the economic and social benefits are taken into account. The advantage of energy conservation measures as the dominant approach to long-run reductions in import dependence is not only that they can effectively achieve this aim but that another result is economizing on resources that can then be employed in other uses. These benefits could be obtained with minimal reductions in consumer welfare and minor increases in production costs.

In the longer run, there are possibilities for a wide range of technological innovations and reorganization of social and institutional arrangements that might be associated with much lower per capita energy consumption levels. The most obvious of these changes include alterations in urban spatial structures and land uses according to energy-minimization criteria, developments in urban transportation (not merely the introduction of traditional public transit facilities, but schemes to promote the development and use of low-cost, standardized vehicles for intraurban use, the promotion of quasi-public car fleets, etc.), and the widespread introduction of solar-heating devices in most new houses. Little is known about the feasibility or acceptability of proposals of this kind. At present, these are questions for research rather than experimentation, not to mention implementation.

As for the supply side of the energy question, different problems arise with each of the main fuels. The oil supply question is complicated by uncertainties about potential resources in Alaska and at offshore locations. The recent Supreme Court decision (March 1975) that the federal government has jurisdiction over offshore oil and gas reserves does not reduce the need for close cooperation between the federal government, the states, and the producing companies in the development of offshore production, especially the question of the location of onshore terminals and refining facilities. In more general terms, the experience of recent years suggests that the oil companies have been conspicuously successful in outmaneuvering the federal government. The importance of energy supplies to the economy is great enough to justify stronger controls and surveillance of the oil companies' operations and behavior. Grounds for concern include queries about deliberate restrictions of supply, especially in offshore fields,[2] diversification into other energy sectors, collusion, and applying pressures to retail sale outlets.

The trouble with the natural gas sector is easy to diagnose, and there now seems to be a wide measure of agreement about it. There has been a natural gas shortage for many years, and this may persist because of limited reserves, either proved or potential. However, the situation has been severely aggravated by the FPC's cumbersome and inefficient policy of price regulation. This has kept the price of gas much below that of other fuels (in relative Btu values), in spite of the fact that the environmental acceptability of natural gas would indicate that its relative price should be higher. The attempts to reduce the discrepancy in 1974 could only be described as "tinkering." The evidence is clear that the shortage has been exacerbated by the chronically low price and its impact on exploration, discoveries, and production. The alternative strategy of complete and rapid deregulation of price controls may not achieve the goals sometimes claimed for it, but it could hardly do worse. Past experience of insecure supplies, reinforced by higher prices, should provide some disincentive to the use of gas, but there may be a case for direct controls to ensure that gas is not used for purposes which could be powered equally well by other fuels.

Coal poses a quite different set of problems, all of which are well known. The critical question is how to expand the use of coal without sacrificing environmental quality standards. A perfect solution may be impossible. Even an acceptable solution may be unlikely until the problem of stack desulfurization is resolved, commercially as well as technically, and until effective strip-mining legislation is not only passed but is fully implemented. Although there are no clear-cut answers to these problems in the short run, it is important that there should be greater coordination of decisions than has been customary in this industry. The federal government is in a position to influence this because of its control of much of the land over western coal fields, its responsibility for safety and environmental legislation, and its capacity to take a broad view of the energy sector as a whole. Probably the soundest guideline is to proceed with caution, without abandoning environmental standards, despoiling the countryside, or radically changing the character of western states, secure in the knowledge that the coal reserves are there if needed. Incremental actions rather than "crash" programs are justified for several reasons: uncertainties about the success and cost of desulfurization; doubts about the long-term viability of coal-fired power stations if the obstacles to widespread adoption of nuclear plants are overcome; the repercussions for transportation (predominantly rail) investments if western coal is to be exploited on a massive scale; the social costs of developing western coal fields, and the federal-state conflicts that might be associated with such development; and the risks of investment in new coal mines (especially for deep coal) because of irreversibility, high capital costs, and the long payoff period.

The electricity utility industry is much more fragmented in the United States than in most other countries. Most of its problems arise from lack of coordination and the large gap between "best practice" and average practice, not only with respect to technology but also social responsibility. In this case the FPC has failed to exercise sufficient control. Most of the difficulties could be resolved by closer cooperation and learning from experience. For instance, there need to be much greater transfers of information about investment decisions; examples include the possibility that too many power companies are investing in semiobsolescent nuclear power plants without regard to the availability and cost of uranium ore supplies or to future technical possibilities, a similar disregard of the implications of too many additional coal-fired plants on low-sulfur coal demand, and the failure to exchange information about future electricity demand. Government action could be useful in coordinating investment decisions, promoting more sensible rate structures, and—possibly most important of all—stimulating research into new technologies, especially to reduce thermal conversion losses. Progress on the latter front is one of the most discouraging aspects of the energy supply situation in recent years. Although this may be improved as a result of the establishment of the Electric Power Research Institute (EPRI), there is also a need for stronger federal support, perhaps via joint sponsorship projects.

The nuclear power question remains an area of bitter controversy, among technologists and social scientists alike. The pessimists have strong arguments in their favor. These relate particularly to the risks of a major reactor accident; a major accident in the transportation of nuclear fuels; nuclear theft by irresponsible or politically motivated groups for the making of bombs or similar threats to society; and the problem of safe disposal of high-level radioactive wastes. On the other hand, it is difficult to evaluate whether these criticisms are overwhelming or merely strong. A zero-risk world is impossible, and estimates of the risks involved are very speculative and subjective. The dangers of a reactor accident depend upon the effectiveness of fail-safe procedures, but the repercussions of an accident could be minimized by greater care in the location of nuclear plants than has been taken hitherto. The transportation and nuclear theft risks could be reduced considerably below present levels (for instance, the recent outcry about the transportation of plutonium oxide by air into big-city airports is a typical example of the scope for improvement). The disposal problem for radioactive wastes has still not been solved satisfactorily (see pp. 100-01, 157), but some of the current lines of investigation could prove effective. Certainly, it is too soon to argue that this problem is not amenable to technological solution.

However serious these qualms may be, they are not going to stop the extension of nuclear power. The number of nuclear power plants planned, under construction, or in operation is already large enough to make it clear that nuclear power is going to be a significant contributor to energy supply in the 1980s. In Europe and Japan, the pace will be even faster, since these countries, lacking the coal reserves and the oil supplies of the United States, have had little option to respond to the OPEC cartel other than by expanding their nuclear programs. The question is not whether or not there is going to be a nuclear energy industry, but how big it will be. In long-term projections of the energy future for the United States, nuclear power assumes an important role in the phase between the decline of oil and the entry of some renewable power source—solar, hydrogen, or even nuclear power itself in the form of fusion. Although the stopgap might be filled with coal, it is arguable that the social costs of using coal (even with effective desulfurization) are greater than in the case of nuclear power. Apart from the grave but improbable risks mentioned above, nuclear energy is very safe in terms of health and injury effects on both the general population and the industry's workers, and is certainly much safer than fossil fuels.

Undoubtedly, there is an understandable but irrational psychological fear about dependence on nuclear power. In part, this derives from its use in bombs and from past fallout scares; in part, it is because the effects of a major disaster are reasonably well appreciated. The problem is not disagreement about the possible effects of such a disaster, but rather the impossibility of estimating the probabilities of its occurrence. There is no way of specifying the trade-off between certain chronic, but low-level, damage from fossil fuel plants and improbable, but possible, cataclysmic damage from nuclear plants. There needs to be greater concern for safety and security than in the past, particularly in view of the expansion of the industry. This may be facilitated by the division of research and regulatory functions with the establishment of ERDA and the NRC. In retrospect, the breeder program can be criticized for its dominance of the energy R and D budget and for its low cost-effectiveness. However, it has gone too far for retracing steps. Indeed, in view of the proliferation of low-efficiency burner reactors, with their high initial fuel requirements, it can be argued that the proper strategy is to prosecute the breeder reactor program as rapidly as possible.[a]

Other fuels—sometimes called the "exotic" fuels—deserve a higher proportion of R and D funds than they have received in the past, and much more encouragement from the federal government. However, many problems remain to be solved, some of them technological, but most of them economic. Taking

---

[a]This remains a controversial point. For a contrary view see Ford Foundation Energy Policy Project, *A Time to Choose* (Cambridge, Mass.: Ballinger, 1974), pp. 202-24.

into account the long lead-times, these fuels are unlikely to make more than a marginal contribution before the mid-1980s. Yet they remain potentially very important—shale and synthetics to guard against depletion of oil and natural gas, solar power and hydrogen as candidates for renewable resources, and tidal and geothermal power as supplementary energy sources at specific locations. There are many ways of stimulating progress in these areas: sponsorship of basic research (e.g. into photovoltaic cell improvements); joint research projects between government and industry; and, above all, demonstration plant projects.

The most critical question raised by any strategy that aims at increasing domestic energy supplies at the expense of imports is the environmental cost. Historically, economic growth in the United States has been given priority over environmental preservation. In recent years, there has been some change of emphasis reflected in legislation such as the National Environmental Policy Act of 1969 and the Clean Air Act and its amendments, as well as in the establishment of new institutions such as the Environmental Protection Agency. These changes resulted in actions to prevent disruption of the environment which were beginning to bite when the "energy crisis" developed. Faced with this new national "problem," the latest of a long line and the most topical, the administration veered in its decisions in favor of resolving the energy-environmental trade-off increasingly in support of energy supply. The veto of the strip-mining bills, the postponement of automobile emission standards, and some relaxation in sulfur oxide emission standards are early examples of this trend. Other serious questions loom ahead, such as how far the administration will go in allowing high-sulfur coal to be burnt, whether compromises will be permitted with water-quality standards, and the fate of western areas containing coal.

The argument here is that the seriousness of the energy situation does not justify wholesale abandonment of environmental standards. The point is important not so much because a year or two's delay in attaining a particular environmental standard would be critical but rather because adhering to environmental objectives is necessary in order to maintain public credibility that the administration takes the environment seriously.

On the other hand, several of the environmental standards have been determined arbitrarily, while the use of the target date approach may result in hurried solutions (e.g. the catalytic converter). The concept of uniform national standards is not very sensible, since the effects of pollutants differ by area according to local economic conditions, climate, and many other factors. The justification for national standards is presumably that they avoid local disputes and pressures for exemptions. However, it would be preferable to allow ad hoc departures from the national standard if the case was strong rather than to abandon the standard as a whole. Environmental considerations should be taken into account in all energy decisions, regardless of whether they are of the scale and type to justify an environmental impact statement or not. A substantial proportion of the energy R and D budget should be devoted to pollution control technologies and other kinds of environmental research. Fuel choice decisions in favor of environmentally unsatisfactory fuels such as high-sulfur coal should be scrutinized most carefully. Power plant location decisions should be treated with more concern for environmental repercussions and effects on nearby populations than has been the case hitherto. It may be inevitable that more emphasis on domestic energy sources must involve a higher degree of environmental disruption than would otherwise have occurred. However, if environmental goals are retained at the forefront of decision, it should be possible to balance energy supply and demand without much additional environmental cost.

Finally, a major question for energy policy is the scope and nature of federal involvement. On any interpretation, the government's role must be extensive. It has traditionally been responsible for the regulation and control of the power industries, it is the owner of a large proportion of energy resources, it is responsible for almost all nuclear power research and for a far from negligible share of research into other fuels, it determines the conditions under which fuels are imported and exported, it influences supply and demand in the energy sector via its tax policies, and it is the setter and guardian of national environmental standards. On the other hand, the US government is much less involved in energy supply than in most other countries, where electricity generation and transmission is a government function and where electricity distribution and gas and coal production may be publicly owned. The problems of coordination are much greater in the American case because the private sector is responsible for all energy supply. Control is difficult in some sectors such as electricity distribution because there are too many small and inefficient concerns, in others such as the oil industry because of the power of a few giant corporations. The multilevel system of government also complicates coordination because the federal government, the states, and local governments all have some areas of overlapping responsibility. Moreover, functions have developed in an unsystematic manner to meet a variety of economic and social objectives not related to energy policy. The result is a high degree of fragmentation of decisions affecting energy policy.

The establishment of the Federal Energy Administration and the Energy Research and Development Administration in 1974 represent important steps to improve the coordination of energy policy and research. Nevertheless, responsibilities remain very diffused, both within the federal government (Department of the Interior, Environmental Protection Agency, etc.) and between different levels of government. For example, the location and siting of power facilities is virtually the sole responsibility of states and municipalities, even though the problem raises issues of national importance. Also, some functions appear to slip through the network of government departments and agencies; energy conservation appears to fall into this category, for example, though the FEA would presumably be responsible in a short-run emergency under the powers granted in the Emergency Petroleum Allocation Act of 1973. Yet, as argued above, energy conservation strategies are perhaps the most important element in formulation of a national energy policy.

Priorities for more effective action by the federal government than in the past include greater spread of information to energy investors, to improve coordination between sectors; a more efficient allocation of resources among competing R and D activities, with particular emphasis on strategic sectors, such as the electricity utilities, and on longer-term and hence riskier possibilities, such as solar power; more sensible federal leasing policies, particularly for coal resources; increasing the effectiveness of regulation of the oil industry, which has remained remarkably free from government supervision compared with other energy producers; investigation of alternative strategies for promoting energy conservation; long-term planning of feasible changes in the fuel mix, and supplying the results to the industry as a whole as a guide to indicative planning; and investigation of the costs and methods of emergency stockpile and reserve capacity schemes.[b]

Conclusion

The existence of a long-term energy crisis is in doubt, and has certainly been exaggerated. There were some uncomfortable moments in 1973 and 1974, readjustments had to be made too quickly to be smooth, and there has been a permanent shift in the relative price ratios between energy and other goods and services. The economic and social repercussions on both the industrial and developing world were severe, and will continue to be felt for many years. On the other hand, the day of energy resource extinction is far off, and it is even doubtful whether the next decade or two will experience a situation of world excess demand for fuels.

This diagnosis does not mean that the impacts of recent events on the United States have been negligible. On the contrary, they were much wider than considering energy as an important resource, industry input, and consumption good would suggest. Perhaps more than any other sequence of events in American history, the painful struggle of the administration in its reluctance to intervene actively in the supply of and demand for fuels (whether in regard to supply stimulants, tax and price instruments, allocations, conservation decrees and other regulations, or participation in research and perhaps eventually in production) brought home to American economic ideologues the inevitable responsibilities of government and the deficiencies of the market in a mixed economy. The shortage of gasoline, and to a lesser extent home heating oil and natural gas, in 1973-74 raised the first self-doubts about the viability, perhaps even about the ethics, of the American way of life—that life style of profligate consumption which is simultaneously the envy and the nightmare of the rest of

---

[b]A preliminary investigation was presented in Federal Energy Administration, *Project Independence Blueprint* (Washington, D.C.: U.S. Government Printing Office, 1974), pp. 369-96. See also Chapter 8.

the world. Just as Vietnam in the 1960s exposed the nerve of American military vulnerability, the energy "crisis" revealed for only the second time in history (1929 was the first) her economic vulnerability. It was a chastening experience. The only important but still unanswered question is: Was it also an educational experience?

Notes

Notes

Chapter 1
Trends and Projections

1. J. Darmstadter, "Energy," pp. 105-49, in Commission on Population Growth and the American Future (R.G. Ridker, ed.), Vol. III, *Population, Resources and the Environment* (Washington, D.C.: US Government Printing Office, 1972).
2. Ford Foundation Energy Policy Project, Final Report, *A Time to Choose: America's Energy Future* (Cambridge, Mass.: Ballinger, 1974), pp. 140-45.
3. Darmstadter.
4. Federal Energy Administration, *Project Independence Blueprint* (Washington, D.C.: US Government Printing Office, November 1974).
5. Darmstadter.
6. A strong case for intrinsic forecasting models is made by W.A. Spivey and W.A. Wecker, "Regional Economic Forecasting: Concepts and Methodology," *Papers, Regional Science Association*, 28 (1972), pp. 257-76.
7. MIT Energy Laboratory Policy Study Group, *Energy Self-Sufficiency: An Economic Evaluation* (Washington, D.C.: American Enterprise Institute, 1974).
8. Ibid., p. 7.

Chapter 2
The Resource Exhaustion Controversy

1. W.S. Jevons, *The Coal Question* (London: Macmillan, 1st edition, 1865; 3rd revised edition, 1906).
2. D.H. Meadows et al., *The Limits to Growth* (New York: Universe Books, 1972).
3. Jevons, p. 2.
4. Ibid., p. 412.
5. Ibid., p. 57.
6. Ibid., p. 7.
7. Ibid., p. 25.
8. Ibid., p. 194.
9. Ibid., pp. 198-99.
10. Ibid., p. 275.
11. Ibid., p. 199.
12. Estimate of E.M. Hull in the *Journal of Science*, 1 (1864), p. 33.
13. Jevons, pp. 274-75.
14. Ibid., p. xxxiv.

15. Ibid., pp. 31-32.
16. Ibid., p. 183.
17. Ibid., p. xliii. These statements illustrate the soundness of the rule—never make a prophecy. Cf. "Uncertainty will for ever render aerial conveyance a commercial impossibility" (ibid., p. 169).
18. Ibid., p. 187.
19. Ibid., p. 190.
20. Ibid., p. 453.
21. Ibid., p. 460.
22. Ibid., p. 456.
23. Ibid., p. 4.
24. W.D. Nordhaus, "Resources as a Constraint on Growth," *American Economic Review*, Papers 64 (1974), p. 23.
25. Meadows et al.
26. Ibid., p. 67.
27. Ibid., pp. 71-78.
28. Ibid., p. 171.
29. Ibid., p. 177.
30. H.D.S. Cole et al., *Models of Doom: A Critique of the Limits to Growth* (New York: Universe Books, 1973), p. 8.
31. J.W. Forrester, *World Dynamics* (Cambridge, Mass.: Wright-Allen Press, 1971).
32. T.W. Oerlemans, M.M.J. Tellings, and H. de Vries, "World Dynamics," *Nature*, Vol. 238 #5362 (August 1972), p. 251.
33. Cole et al., p. 86.
34. H.G. Barnett and C.W. Morse, *Scarcity and Growth: The Economics of Natural Resource Availability* (Baltimore: Johns Hopkins Press for Resources for the Future, 1963).
35. Ibid., p. 7.
36. Ibid., p. 10.
37. Ibid., p. 11.
38. Ibid., p. 249.
39. Nordhaus.
40. W.D. Nordhaus, "The Allocation of Energy Resources," *Brookings Papers on Economic Activity*, 3 (1973), pp. 529-70. See also Nordhaus (1974).
41. Nordhaus (1973), p. 570.
42. Nordhaus (1974).

Chapter 3
The Theory of Exhaustible Resources

1. L.C. Gray, "Rent under the Assumption of Exhaustibility," *Quarterly Journal of Economics*, 28 (1914), pp. 66-89; H. Hotelling "The Economics of

Exhaustible Resources," *Journal of Political Economy*, 39 (1931), pp. 137-75; A.D. Scott, "The Theory of the Mine under Conditions of Certainty," pp. 25-62, in M. Gaffney (ed.), *Extractive Resources and Taxation* (Madison: University of Wisconsin Press, 1967); R.L. Gordon, "A Reinterpretation of the Pure Theory of Exhaustion," *Journal of Political Economy*, 75 (1967), pp. 274-86; P.G. Bradley, "Increasing Scarcity: The Case of Energy Resources," *American Economic Review*, Papers 63 (1973), pp. 119-25; W.D. Nordhaus, "The Allocation of Energy Resources," *Brookings Papers on Economic Activity*, 3 (1973), pp. 529-70; R.M. Solow, "The Economics of Resources and the Resources of Economics," *American Economic Review*, Papers 64 (1974), pp. 1-14; F.E. Banks, "A Note on Some Theoretical Issues of Resource Depletion," *Journal of Economic Theory*, 9 (1974), pp. 238-43; and W.D. Schulze, "The Optimal Use of Non-Renewable Resources: The Theory of Extraction," *Journal of Environmental Economics and Management*, 1 (1974), pp. 53-73. The analysis adopted here is closer to that of Gordon.

2. For an example see Schulze.

3. Scott.

4. See Schulze, pp. 54-58.

5. Solow.

6. M. Gaffney, "Editor's Conclusion," pp. 333-419, in Gaffney.

7. O.C. Herfindahl, "Depletion and Economic Theory," pp. 63-90, in Gaffney.

8. H.G. Barnett and C.W. Morse, *Scarcity and Growth: The Economics of Natural Resource Availability* (Baltimore: Johns Hopkins Press for Resources for the Future, 1963).

9. Ibid.

10. Nordhaus (1973).

11. For further extensions see V.L. Smith, "An Optimistic Theory of Exhaustible Resources," *Journal of Economic Theory*, 9 (1974), pp. 384-96.

12. A.D. Scott, *Natural Resources: The Economics of Conservation* (Toronto: University of Toronto Press, 1955) and Scott (1967).

13. K.J. Arrow and R.C. Lind, "Uncertainty and the Evaluation of Public Investment Decisions," *American Economic Review*, 60 (1970), pp. 364-78.

14. W.J. Baumol, "On the Social Rate of Discount," *American Economic Review*, 58 (1968), pp. 788-802.

15. Solow, p. 12.

16. F.P. Ramsey, "A Mathematical Theory of Saving," *Economic Journal*, 38 (1928), pp. 543-59.

17. G.M. Heal and P. Dasgupta, "The Optimal Depletion of Exhaustible Resources," *Review of Economic Studies*, forthcoming.

18. Solow.

19. R.L. Gordon, "Conservation and the Theory of Exhaustible Resources," *Canadian Journal of Economics and Political Science*, 31 (1966), pp. 319-26.

20. Ibid., p. 326.

21. S.A. Marglin, "The Social Rate of Discount and the Optimal Rate of Investment," *Quarterly Journal of Economics*, 77 (1963), pp. 95-111.

Chapter 4
Oil

1. P. Davidson, "Public Policy Problems of a Domestic Crude Oil Industry," *American Economic Review*, 53 (1963), p. 92.

2. S.L. McDonald, *Petroleum Conservation in the United States: An Economic Analysis* (Baltimore: Johns Hopkins Press, 1971).

3. Davidson, p. 106.

4. G.P. Jenkins and B.D. Wright, "Taxation of Income of Multinational Corporations: The Case of the United States Petroleum Industry," *Review of Economics and Statistics*, 57 (1975), pp. 1-11.

5. T. Duchesneau, *Competition in the Energy Industry* (Cambridge, Mass.: Ballinger, 1974).

6. B.I. Oppenheimer, *Oil and the Congressional Process* (Lexington, Mass.: D.C. Heath, Lexington Books, 1974).

7. Cabinet Task Force on Oil Import Control, *The Oil Import Question* (Washington, D.C.: US Government Printing Office, 1970), p. 98.

8. W.D. Nordhaus, "The 1974 Report of the President's Council of Economic Advisers: Energy in the Economic Report," *American Economic Review*, 64 (1974), p. 560.

9. M.A. Adelman, "Politics, Economics, and World Oil," *American Economic Review*, Papers 64 (1974), pp. 58-67.

Chapter 5
Coal, Gas and Electricity

1. C.F. Wen and R.T. Newcomb, "The Potential of Coal-Based Fuel and Energy Complexes: A Regional Appraisal" (National Science Foundation–RANN Report, January 1974).

2. Federal Energy Administration, *Project Independence Blueprint* (Washington, D.C.: US Government Printing Office, 1974), p. 9.

3. Federal Power Commission, *The Potential for Conversion of Oil-Fired and Gas-Fired Electric Generating Units to Use of Coal* (Washington, D.C. FPC, 1973).

4. PIB, p. 108, and MIT Energy Policy Laboratory Study Group, *Energy Self-Sufficiency: An Economic Evaluation* (Washington, D.C.: American Enterprise Institute, 1974), p. 42.

5. Ford Foundation Energy Policy Project, Final Report, *A Time to Choose: America's Energy Future* (Cambridge, Mass.: Ballinger, 1974), p. 296.

6. P.W. MacAvoy and R.S. Pindyck, "Alternative Regulatory Policies for Dealing with the Natural Gas Shortage," *Bell Journal of Economics and Management Science*, 4 (1973), pp. 454-98.

7. E.W. Erickson and R.M. Spann, "Supply Response in a Regulated Industry: The Case of Natural Gas," *Bell Journal of Economics and Management Science*, 2 (1971), pp. 94-121.

8. National Petroleum Council, *Report on United States Energy Outlook* (Washington: National Petroleum Council, 1972).

9. R.B. Helms, *Natural Gas Regulation: An Evaluation of FPC Price Controls* (Washington, D.C.: American Enterprise Institute, 1974), and S.G. Breyer and P.W. MacAvoy, *Energy Regulation by the Federal Power Commission* (Washington, D.C.: Brookings Institution, 1974).

10. J.D. Khazzoom, "The FPC Staff's Econometric Model of Natural Gas Supply in the United States," *Bell Journal of Economics and Management Science*, 2 (1971), pp. 51-93.

11. See also Foster Associates, *Energy Prices, 1960-73* (Cambridge, Mass.: Ford Foundation Energy Policy Project, Ballinger, 1974).

12. C.E. Olson, *Cost Considerations for Efficient Electricity Supply* (East Lansing: Public Utilities Papers, Michigan State University, 1970). Earlier studies include R. Komiya, "Technological Progress and the Production Function in the United States Steam Power Industry," *Review of Economics and Statistics*, 44 (1962), pp. 156-66, and S. Ling, *Economies of Scale in the Steam-Electric Power Generating Industry* (Amsterdam: North Holland, 1964).

Chapter 6
Nuclear Power

1. W.D. Nordhaus, "Resources as a Constraint on Growth," *American Economic Review*, Papers 64 (1974), pp. 22-26.

2. For a negative view see T.B. Cochran, *The Liquid Metal Fast Breeder Reactor: An Economic and Environmental Critique* (Baltimore: Johns Hopkins Press, for Resources for the Future, 1973).

3. M. King Hubbert, "Energy Resources," p. 230, in National Academy of Sciences—National Research Council, Committee on Resources and Man, *Resources and Man* (San Francisco: W.H. Freeman & Co., 1969).

4. W.D. Nordhaus, "The Allocation of Energy Resources," *Brookings Papers on Economic Activity*, 3 (1973), pp. 529-70.

5. A.B. Lovins, "Energy Resources," United Nations Symposium on Population, Resources and the Environment, Stockholm, September-October 1973.

6. Nordhaus (1973), unpublished appendix.

7. National Petroleum Council, *Report on United States Energy Outlook* (Washington: National Petroleum Council, 1972).

8. Lovins.

9. F.P. Baranowski, "Uranium Supplies, Costs and AEC Resource Evaluation Program," AEC Oak Ridge Laboratory, Tennessee, November 1974.

10. L.B. Lave and L.C. Freeburg, "Health Costs to the Consumer per Megawatt-Hour of Electricity," in A.J. Finkel (ed.), *Energy, the Environment and Human Health* (Acton, Mass.: for the American Medical Association, Publishing Sciences Group Inc., 1974), p. 220.

11. M. Willrich and T.B. Taylor, *Nuclear Theft: Risks and Safeguards* (Cambridge, Mass.: Ballinger, 1974).

12. Ibid., p. 173.

13. *New York Sunday Times*, November 10, 1974.

Chapter 7
Other Fuels

1. MIT Energy Laboratory Policy Study Group, *Energy Self-Sufficiency: An Economic Evaluation* (Washington, D.C.: American Enterprise Institute, 1974).

2. R.A. Tybout and G.O.G. Löf, "Solar House Heating," *Natural Resources Journal*, 10 (1970), pp. 268-326. Although this study is dated, conventional fuel costs have risen in the past five years faster than those of solar collector devices, and with large-scale commercial production the scope for economies is substantial.

3. D.W. White, "Geothermal Resources," in P. Kruger and C. Otte (eds.), *Geothermal Energy: Resources, Production, Stimulation* (Stanford: Stanford UP, 1973).

4. R.A. Geyer, "Energy from the Oceans," pp. 94-104, in T.S. English (ed.), *Ocean Resources and Public Policy* (Seattle: Washington UP, 1973).

5. D.P. Gregory, "The Hydrogen Economy," *Scientific American*, 228 (1973), pp. 13-22.

Chapter 8
OPEC, the United States, and World Oil

1. See Federal Energy Administration, *Project Independence Blueprint* (Washington, D.C.: US Government Printing Office, 1974), Appendix A5, pp. 283-304.

2. G.W. Gruver and R.S. Thorn, "The Optimal Rate of Depletion of an Exhaustible Natural Resource for a Small Country" (University of Pittsburgh, Department of Economics, mimeo, 1974).

3. M.A. Adelman, "Politics, Economics and World Oil," *American Economic Review*, Papers 64 (1974), pp. 58-67.

4. J.W. McKie, "The Political Economy of World Petroleum," *American Economic Review*, Papers 64 (1974), pp. 51-57.

5. P. O'Dell, *Oil and World Power: Background to the Oil Crisis* (Harmondsworth, Middlesex: Penguin, 1974), pp. 203-12.

6. W.D. Nordhaus, "The 1974 Report of the President's Council of Economic Advisers: Energy in the Economic Report," *American Economic Review*, 64 (1974), pp. 558-65.

7. Brookings Institution, *Energy and US Foreign Policy* (Report to Ford Foundation Energy Policy Project, June 1974).

8. MIT Energy Laboratory Policy Study Group, *Energy Self-Sufficiency: An Economic Evaluation* (Washington, D.C.: American Enterprise Institute, 1974), pp. 69-71.

9. PIB, pp. 369-96.

Chapter 9
Demand Restraint

1. Department of Housing and Urban Development (HUD), *Residential Energy Consumption. Single Family Housing. Final Report* (Washington, D.C.: Department of Housing and Urban Development, Contract #H-1654, March 1973).

2. Ibid.

3. B. Pushkarev, "Energy in the New York Region," pp. 13-23, in R.H. Connery and R.S. Gilmour (eds.), *The National Energy Problem* (Lexington, Mass.: D.C. Heath, Lexington Books, 1974).

4. G.A. Lincoln, "Energy Conservation," *Science*, April 13, 1973.

5. E. Hirst, *Energy Consumption for Transportation in the United States* (Oak Ridge National Laboratory, Tennessee, March 1972); and E. Hirst, "Energy-Intensiveness of Transportation," *Transportation Engineering Journal*, 99 (1973), p. 111.

6. Regional Plan Association, *Transportation and Economic Opportunity: A Report to the Transportation Administration of the City of New York* (New York, 1973).

7. Environmental Protection Agency, *The Potential for Energy Conservation* (Washington, D.C.: Office of Emergency Preparedness, 1972).

8. S. Wildhorn, B.K. Burright, J. Enns, and T.F. Kirkwood, *How to Save Gasoline: Public Policy Alternatives to the Automobile* (Santa Monica: RAND Corporation, 1974).

9. Regional Plan Association.

10. R. Halvorsen, "Residential Demand for Electric Energy," *Review of Economics and Statistics*, 57 (1975), pp. 12-18.

Chapter 10
Energy and the Environment

1. For a comprehensive analysis, see D.L. Scott, *Pollution in the Electric Power Industry: Its Control and Costs* (Lexington, Mass.: D.C. Heath, Lexington Books, 1973).

2. N. Fabricant and R.M. Hallman, *Toward a Rational Power Policy: Energy, Politics and Pollution* (New York: for the Environmental Protection Administration of New York, G. Braziller, 1972).

3. C.L. Wilson and W.H. Matthews (eds.), *Man's Impact on the Global Environment* (Cambridge, Mass.: Study of Critical Environmental Problems, MIT, 1970).

4. W.D. Nordhaus, "Resources as a Constraint on Growth," *American Economic Review*, Papers 64 (1974), p. 26.

5. D.W. Ducsik (ed.), *Power, Pollution and Public Policy* (Cambridge, Mass.: MIT Press, 1971).

6. R.P. Ouellette, "Energy and Environmental Quality," pp. 170-82, in R.H. Connery and R.S. Gilmour (eds.), *The National Energy Problem* (Lexington, Mass.: D.C. Heath, Lexington Books, 1974).

7. For a case study see W.R. Ahern, Jr., *Oil and the Outer Coastal Shelf: The Georges Bank Case* (Cambridge, Mass.: Ballinger, 1973).

8. C.J. Cicchetti, *Alaskan Oil: Alternative Routes and Markets* (Baltimore: Johns Hopkins Press for Resources for the Future, 1972).

9. Ibid.

Chapter 11
Some Policy Considerations

1. J.E. Tilton, *US Energy R and D Policy* (Washington, D.C.: Resources for the Future, 1974).

2. G.M. Brannon, *Energy Taxes and Subsidies* (Cambridge, Mass.: Ballinger, 1974). See also G.M. Brannon (ed.), *Studies in Energy Tax Policy* (Cambridge, Mass.: Ballinger, 1974).

3. T. Duchesneau, *Competition in the Energy Industry* (Cambridge, Mass.: Ballinger, 1974).

4. S.G. Breyer and P.W. MacAvoy, *Energy Regulation by the Federal Power Commission* (Washington, D.C.: Brookings Institution, 1974), pp. 14-15.

5. MIT Energy Laboratory Policy Study Group, *Energy Self-Sufficiency: An Economic Evaluation* (Washington, D.C.: American Enterprise Institute, 1974).

6. Ford Foundation Energy Policy Project, *A Time to Choose: America's Energy Future* (Cambridge, Mass.: Ballinger, 1974).

7. Federal Energy Administration, *Project Independence Blueprint* (Washington, D.C.: US Government Printing Office, 1974).

8. PIB, p. 45.

9. B.M. Hannon, "An Energy Standard of Value," *Annals of the American Academy of Political and Social Science,* #410 (1973), pp. 139-153.

Chapter 12
Toward a National Energy Policy

1. Federal Energy Administration, *Project Independence Blueprint* (Washington, D.C.: US Government Printing Office, 1974), p. 318.

2. P. Davidson et al., *Offshore Drilling* (Washington, D.C.: Brookings Institution, 1974).

Bibliography

Bibliography

Adelman, M.A. "Efficiency in Resource Use in Crude Petroleum," *Southern Economic Journal*, 31 (1964), pp. 101-22.

―――. *The World Petroleum Market* (Baltimore: Johns Hopkins Press for Resources for the Future, 1972).

―――. "Is the Oil Shortage Real? Oil Companies as OPEC Tax Collectors," *Foreign Policy*, #9 (1972), pp. 69-107.

―――. "Politics, Economics and World Oil," *American Economic Review*, Papers 64 (1974), pp. 58-67.

Agria, S.R. "Special Tax Treatment of Mineral Industries," pp. 77-122, in A.C. Harberger and M.J. Bailey (eds.), *The Taxation of Income from Capital* (Washington, D.C.: Brookings Institution, 1969).

Ahern, Jr., W.R. *Oil and the Outer Coastal Shelf: The Georges Bank Case* (Cambridge, Mass.: Ballinger, 1973).

Akins, J. "The Oil Crisis: This Time the Wolf Is Here," *Foreign Affairs*, 50 (1972), pp. 462-90.

American Gas Association, *Gas Facts* (New York: American Gas Association, annual).

American Gas Association, American Petroleum Institute and Canadian Petroleum Association, *Reserves of Crude Oil, Natural Gas Liquids and Natural Gas in the United States and Canada and United States Productive Capacity* (New York, May 1973).

American Petroleum Institute, *Petroleum Facts and Figures* (New York: annual).

Amuzegar, J. "The Oil Story: Facts, Fiction and Fair Play," *Foreign Affairs*, 51 (1973), pp. 676-89.

Anderson, K. "Optimal Growth When the Stock of Resources Is Finite and Depletable," *Journal of Economic Theory*, 4 (1972), pp. 256-67.

Arrow, K.J., and R.C. Lind. "Uncertainty and the Evaluation of Public Investment Decisions," *American Economic Review* 60 (1970), pp. 364-78.

Banks, F.E. "A Note on Some Theoretical Issues of Resource Depletion," *Journal of Economic Theory*, 9 (1974), pp. 238-43.

Baranowski, F.P. "Uranium Supplies, Costs and AEC Resource Evaluation Program" (AEC Oak Ridge Laboratory, Tennessee, November 1974).

Barnett, H.G., and C.W. Morse. *Scarcity and Growth: The Economics of Natural Resource Availability* (Baltimore: Johns Hopkins Press for Resources for the Future, 1963).

Baumol, W.J. "On the Social Rate of Discount," *American Economic Review*, 58 (1968), pp. 788-802.

Bernshtein, L.B. *Tidal Energy for Electric Power Plants* (Jerusalem: Israel Program for Scientific Translations, 1965).

Boesch, D.F., C.H. Hershner, and J.H. Milgram. *Oil Spills and the Marine Environment* (Cambridge, Mass.: Ballinger, 1974).

Bradley, P.G. *The Economics of Crude Petroleum Production* (Amsterdam: North Holland, 1967).

_____. "Increasing Scarcity: The Case of Energy Resources," *American Economic Review*, Papers 63 (1973), pp. 119-25.

Brannon, G.M. *Energy Taxes and Subsidies* (Cambridge, Mass.: Ballinger, 1974).

_____ (ed.). *Studies in Energy Tax Policy* (Cambridge, Mass.: Ballinger, 1974).

Breyer, S.G., and P.W. MacAvoy. *Energy Regulation by the Federal Power Commission* (Washington, D.C.: Brookings Institution, 1974).

Brookings Institution. *Energy and United States Foreign Policy* (Report to the Ford Foundation Energy Policy Project, June 1974).

Brooks, D.B. (ed.). *Resource Economics: Selected Works of O.C. Herfindahl* (Baltimore: Johns Hopkins Press, 1974).

Buchanan, J.M., and T.N. Tideman. "Gasoline Rationing and Market Pricing: Public Choice in Political Economy," *Atlantic Economic Journal*, 2 (1974).

Burrows, J.C., and T.S. Domencich. *An Analysis of the US Oil Import Quota* (Lexington, Mass.: D.C. Heath, Lexington Books, 1970).

Carter, A.P. "Energy, Environment and Economic Growth," *Bell Journal of Economics and Management Science*, 5 (1974), pp. 578-92.

Carter, A.P. "Application of Input-Output Analysis to Energy Problems," *Science*, April 19, 1974, pp. 325-29.

Cicchetti, C.J. *Alaskan Oil: Alternative Routes and Markets* (Baltimore: Johns Hopkins Press for Resources for the Future, 1972).

Clark, W. *Energy for Survival: The Alternative to Extinction* (New York: Doubleday, Anchor Books, 1974).

Cochran, T.B. *The Liquid Metal Fast Breeder Reactor: An Economic and Environmental Critique* (Baltimore: Johns Hopkins Press for Resources for the Future, 1973).

Cole, H.S.D. et al. *Models of Doom: A Critique of the Limits to Growth* (New York: Universe Books, 1973).

Connery, R.H., and R.S. Gilmour (eds.). *The National Energy Problem* (Lexington, Mass.: D.C. Heath, Lexington Books, 1974).

Council on Economic Priorities. *The Price of Power, Electric Utilities and the Environment* (Washington, D.C.: 1972).

Cummings, R.G. "Some Extensions of the Economic Theory of Exhaustible Resources," *Western Economic Journal*, 7 (1969), pp. 201-10.

Cummings, R.G. et al. "Energy Commodities and Natural Resource Exploitation," *Southern Economic Journal*, 41 (1975), pp. 382-94.

Darmstadter, J. "Energy," pp. 105-49 in R.G. Ridker (ed.) for the Commission on Population Growth and the American Future, Vol. III, *Population, Resources and the Environment* (Washington, D.C.: US Government Printing Office, 1972).

_____. *Regional Energy Consumption* (Washington, D.C.: Resources for the Future and Regional Plan Association, 1973).

_____ et al. *Energy in the World Economy* (Baltimore: Johns Hopkins Press, 1972).

Davidson, P. "Public Policy Problems of a Domestic Crude Oil Industry," *American Economic Review*, 53 (1963), pp. 85-108.

_____ et al. *Offshore Drilling* (Washington, D.C.: Brookings Institution, 1974).

Doctor, R.D., et al. *California's Electric Quandary: Slowing the Growth Rate* (Santa Monica: RAND Corporation, 1972).

Duchesneau, T. *Competition in the Energy Industry* (Cambridge, Mass.: Ballinger, 1974).

Ducsik, D.W. (ed.). *Power, Pollution and Public Policy* (Cambridge, Mass.: MIT Press, 1971).

Dupree, W.A., and J.A. West. *United States Energy Through the Year 2000* (US Department of the Interior, December 1972).

Edison Electric Institute. *Statistical Yearbook of the Electricity Utility Industry* (New York: annual).

Efford, I.E., and B.M. Smith (eds.), *Energy and the Environment* (Vancouver: Institute of Resource Ecology, University of British Columbia, 1972).

Erickson, E.W., and R.M. Spann. "Supply Response in a Regulated Industry: The Case of Natural Gas," *Bell Journal of Economics and Management Science*, 2 (1971), pp. 94-121.

Erickson, E.W. and L. Waverman (eds.). *The Energy Question: An International Failure of Policy*, 2 vols. (Toronto: University of Toronto Press, 1974).

Fabricant, N., and R.M. Hallman. *Toward a Rational Power Policy: Energy, Politics and Pollution* (New York: For the Environmental Protection Administration of New York, G. Braziller, 1972).

Fisher, F.M. *Supply and Cost in the US Petroleum Industry* (Baltimore: Johns Hopkins Press, 1964).

Ford Foundation Energy Policy Project. *A Time to Choose: America's Energy Future* (Cambridge, Mass.: Ballinger, 1974).

Forrester, J.W. *World Dynamics* (Cambridge, Mass.: Wright-Allen Press, 1971).

Foster Associates. *Energy Prices, 1960-73* (Cambridge, Mass.: Ford Foundation Energy Policy Project, Ballinger, 1974).

Frank, H., and D. Well. "US Oil Imports: Implications for the Balance of Payments," *Natural Resources Journal*, 13 (1973), pp. 431-47.

Freeman, S.D. *Energy: The New Era* (New York: For Twentieth Century Fund, Walker & Co., 1974).

Gaffney, M. (ed.). *Extractive Resources and Taxation* (Madison: University of Wisconsin Press, 1967).

Georgescu-Roegen, N. "Energy and Economic Myths," *Southern Economic Journal*, 41 (1975), pp. 347-81.

Gerwig, R.W. "Natural Gas Production: A Study of Costs of Regulation," *Journal of Law and Economics*, 5 (1962), pp. 69-92.

Geyer, R.A. "Energy From the Oceans," pp. 94-104, in T.S. English (ed.), *Ocean Resources and Public Policy* (Seattle: Washington University Press, 1973).

Goldsmith, O.S. "Market Allocation of Exhaustive Resources," *Journal of Political Economy*, 82 (1974), pp. 1035-40.

Gordon, R.L. "Conservation and the Theory of Exhaustible Resources," *Canadian Journal of Economics and Political Science*, 31 (1966), pp. 319-26.

———. "A Reinterpretation of the Pure Theory of Exhaustion," *Journal of Political Economy*, 75 (1967), pp. 274-86.

Gray, L.C. "Rent under the Assumption of Exhaustibility," *Quarterly Journal of Economics*, 28 (1914), pp. 66-89.

Gregory, D.P. "The Hydrogen Economy," *Scientific American*, 228 (1973), pp. 13-22.

Griffin, J.M. "An Econometric Evaluation of Sulfur Taxes," *Journal of Political Economy*, 82 (1974), pp. 669-88.

Griffin, J.M. "The Effects of Higher Prices on Electricity Consumption," *Bell Journal of Economics and Management Science*, 5 (1974), pp. 515-39.

Gruver, G.W., and R.S. Thorn. "The Optimal Rate of Depletion of an Exhaustible Natural Resource for a Small Country," (University of Pittsburgh, Department of Economics, mimeo, 1974).

Guyol, N.G. *The World Electric Power Industry* (Berkeley: University of California Press, 1969).

Halvorsen, R. "Residential Demand for Electric Energy," *Review of Economics and Statistics*, 57 (1975), pp. 12-18.

Hammond, A., W. Metz, and T. Maugh. *Energy and the Future* (Washington, D.C.: American Association for the Advancement of Science, 1973).

Hannon, B.M. "An Energy Standard of Value," *Annals of the American Academy of Political and Social Science,* #410 (1973), pp. 139-53.

Harbeson, R.W. "Transport Policy Implications of Energy Shortages," *Land Economics*, 50 (1974), pp. 387-96.

Hawkins, C.A. *The Field Price Regulation of Natural Gas* (Tallahassee: Florida State University Press, 1969).

Heal, G.M., and P. Dasgupta. "The Optimal Depletion of Exhaustible Resources," *Review of Economic Studies*, forthcoming.

Helms, R.B. *Natural Gas Regulation: An Evaluation of FPC Price Controls* (Washington, D.C.: American Enterprise Institute, 1974).

Herfindahl, O.C. "Goals and Standards of Performance for the Conservation of Minerals," *Natural Resources Journal*, 3 (1963), pp. 78-97.

Herfindahl, O.C. "Depletion and Economic Theory," pp. 63-90, in M. Gaffney (ed.), *Extractive Resources and Taxation* (Madison: University of Wisconsin Press, 1967).

Herfindahl, O.C., and A.V. Kneese. *Economic Theory of Natural Resources* (Columbus, Ohio: C.E. Merrill, 1974).

Hirshleifer, J., and D.L. Shapiro. "The Treatment of Risk and Uncertainty," *Quarterly Journal of Economics*, 77 (1963), pp. 95-111.

Hirst, E. *Energy Consumption for Transportation in the United States* (Oak Ridge National Laboratory, Tennessee, March 1972).

_____. "Energy-Intensiveness of Transportation," *Transportation Engineering Journal*, 99 (1973).

Holdren, J., and P. Herrera. *Energy* (New York: Sierra Club Books, 1972).

Hotelling, H. "The Economics of Exhaustible Resources," *Journal of Political Economy*, 39 (1931), pp. 137-75.

Hottel, H.C., and J.B. Howard. *New Energy Technology: Some Facts and Assessments* (Cambridge, Mass.: MIT Press, 1971).

Hubbert, M. King. "Energy Resources," pp. 157-242, in National Academy of Sciences-National Research Council, Committee on Resources and Man, *Resources and Man* (San Francisco: W.H. Freeman & Co., 1969).

Hudson, E.A. and D.W. Jorgenson. "US Energy Policy and Economic Growth, 1975-2000." *Bell Journal of Economics and Management Science*, 5 (1974), pp. 461-514.

Hull, E.M. *Journal of Science*, 1 (1864), p. 33.

Inglis, D.R. *Nuclear Energy: Its Physics and its Social Challenge* (Reading, Mass.: Addison-Wesley, 1973).

Issawi, C. *Oil, the Middle East and the World* (New York: The Library Press, 1972).

Jacoby, N.H. *Multinational Oil: A Study in Industrial Dynamics* (New York: Macmillan, 1974).

_____ and J. Steinbrunner. "Salvaging the Federal Attempt to Control Auto Pollution," *Public Policy*, 21 (1973), pp. 1-48.

Jenkins, G.P., and B.D. Wright. "Taxation of Income of Multinational Corporations: The Case of the United States Petroleum Industry," *Review of Economics and Statistics*, 57 (1975), pp. 1-11.

Jevons, W.S. *The Coal Question* (London: Macmillan, 1st ed. 1865, 3rd ed. 1906).

Kahn, A.E. "The Depletion Allowance in the Context of Cartelization," *American Economic Review*, 54 (1964), pp. 286-314.

Kennedy, M. "An Economic Model of the World Oil Market," *Bell Journal of Economics and Management Science*, 5 (1974), pp. 540-77.

Khazzoom, J.D. "The FPC Staff's Econometric Model of Natural Gas Supply in the United States," *Bell Journal of Economics and Management Science*, 2 (1971), pp. 51-93.

Kitch, E.W. "Regulation and the Field Market for Natural Gas," *Journal of Law and Economics*, 11 (1968), pp. 243-80.

Komiya, R. "Technological Progress and the Production Function in the United States Steam Power Industry," *Review of Economics and Statistics*, 44 (1962), pp. 156-66.

Koopmans, T.C. "Some Observations on 'Optimal' Economic Growth and Exhaustible Resources," pp. 239-56, in H.C. Bos (ed.), *Economic Structure and Development* (Amsterdam: North Holland, 1973).

Kruger, P., and C. Otte (eds). *Geothermal Energy: Resources, Production, Stimulation* (Stanford: Stanford University Press, 1973).

Krutilla, J.V. "Conservation Reconsidered," *American Economic Review*, 57 (1967), pp. 777-86.

Kuller, R.G., and R.G. Cummings. "An Economic Model of Production and Investment for Petroleum Reservoirs," *American Economic Review*, 64 (1974), pp. 66-79.

Landsberg, H.H., and S.H. Schurr. *Energy in the United States: Sources, Uses and Policy Issues* (New York: Random House, 1968).

Landsberg, H.H., J.J. Schanz, Jr., S.H. Schurr, and G.P. Thompson (eds.), *Energy and the Social Sciences: An Examination of Research Needs* (Washington, D.C.: Resources for the Future, 1974).

Lapp, R.E. *The Logarithmic Century* (Englewood Cliffs, N.J.: Prentice Hall, 1973).

Lave, L.B., and L.C. Freeburg. "Health Costs to the Consumer per Megawatt-Hour of Electricity," pp. 209-27, in A.J. Finkel (ed.), *Energy, the Environment and Human Health* (Acton, Mass.: Publishing Sciences Group Inc. for the American Medical Association, 1974).

Lawrence, R.M., and N.I. Wengert (eds.), *The Energy Crisis: Reality or Myth* (Philadelphia: American Academy of Political and Social Science, 1973).

Lincoln, G.A. "Energy Conservation," *Science*, 180, #4082 (1973), pp. 155-62.

Ling, S. *Economies of Scale in the Steam-Electric Power Generating Industry* (Amsterdam: North Holland, 1964).

Löf, G.O.G. "Solar Energy: An Infinite Source of Clean Energy," *Annals of the American Academy of Political and Social Science,* #410 (1973), pp. 52-64.

Lovins, A.B. "Energy Resources" (United Nations Symposium on Population, Resources and the Environment, Stockholm, September-October 1973).

MacAvoy, P.W. "The Effectiveness of the Federal Power Commission," *Bell Journal of Economics and Management Science*, 1 (1970), pp. 271-303.

_____. "The Formal Work-Product of the Federal Power Commissioners," *Bell Journal of Economics and Management Science*, 2 (1971), pp. 379-95.

_____ and R.S. Pindyck. "Alternative Regulatory Policies for Dealing with the Natural Gas Shortage," *Bell Journal of Economics and Managment Science*, 4 (1973), pp. 454-98.

McCloskey, M. "The Energy Crisis: The Issues and a Proposed Response," *Environmental Affairs*, 1 (1971), pp. 587-605.

McDonald, S.L. *Petroleum Conservation in the United States: An Economic Analysis* (Baltimore: Johns Hopkins Press, 1971).

McKie, J.W. "Market Structure and Uncertainty in Oil and Gas Regulation," *Quarterly Journal of Economics*, 74 (1960), pp. 543-71.

_____. "The Political Economy of World Petroleum," *American Economic Review*, Papers 64 (1974), pp. 51-57.

_____ and S.L. McDonald. "Petroleum Conservation in Theory and Practice," *Quarterly Journal of Economics*, 76 (1962), pp. 98-121.

Mancke, R.B. *The Failure of US Energy Policy* (New York: Columbia University Press, 1974).

Marglin, S.A. "The Social Rate of Discount and the Optimal Rate of Investment," *Quarterly Journal of Economics*, 77 (1963), pp. 95-111.

Massachusetts Institute of Technology Energy Laboratory Policy Study Group. *Energy Self-Sufficiency: An Economic Evaluation* (Washington, D.C.: American Enterprise Institute, 1974).

Mead, W. "The Cost of Storing Oil" (Testimony before the Committee on Interior and Insular Affairs, US Senate, May 30, 1973).

_____ and P. Sorensen. "A National Defense Petroleum Reserve Alternative to Oil Import Quotas," *Land Economics*, 47 (1971), pp. 211-24.

Meadows, D.H. et al. *The Limits to Growth* (New York: Universe Books, 1972).

Miller, E. "Some Implications of Land Ownership Patterns for Petroleum Policy," *Land Economics*, 46 (1973), pp. 413-23.

Miller, R.L. *The Economics of Energy* (New York: Morrow, 1974).

National Academy of Sciences and National Academy of Engineering. *The Rehabilitation Potential of Western Coal Lands* (Cambridge, Mass.: Ballinger, 1974).

National Petroleum Council. *Report on United States Energy Outlook* (Washington, D.C.: National Petroleum Council, 1972).

_____. *An Initial Appraisal by the Oil Shale Task Group* (Washington, D.C.: NPC, 1972).

Nordhaus, W.D. "The Allocation of Economic Resources," *Brookings Papers on Economic Activity*, 3 (1973), pp. 529-70.

_____. "Resources as a Constraint on Growth," *American Economic Review*, Papers 64 (1974), pp. 22-26.

_____. "The 1974 Report of the President's Council of Economic Advisers: Energy in the Economic Report," *American Economic Review*, 64 (1974), pp. 558-65.

O'Dell, P. *Oil and World Power: Background to the Oil Crisis* (Harmondsworth, Middlesex: Penguin, 1974).

Oerlemans, T.W., M.M.J. Tellings, and H. De Vries. "World Dynamics," *Nature*, Vol. 238, #5362, August 1972, p. 251.

Olson, C.E. *Cost Considerations for Efficient Electricity Supply* (East Lansing: Public Utilities Papers, Michigan State University Press, 1970).

Olson, M., and H.H. Landsberg (eds.). *The No-Growth Society* (New York: Norton, 1973).

Oppenheimer, B.I. *Oil and the Congressional Process* (Lexington, Mass.: D.C. Heath, Lexington Books, 1974).

Organization for Economic Cooperation and Development (OECD). *Oil: The Present Situation and Future Prospects* (Paris: OECD, 1973).

Ouellette, R.P. "Energy and Environmental Quality," pp. 170-82, in R.H. Connery and R.S. Gilmour (eds.), *The National Energy Problem* (Lexington, Mass.: D.C. Heath, Lexington Books, 1974).

Pearce, P.H. (ed.), *The Mackenzie Pipeline: Arctic Gas and Canadian Energy Policy* (Toronto: McClelland and Stewart, 1974).

Pikl, Jr., I.J. (ed.), *Public Policy and the Future of the Petroleum Industry* (Laramie, Wyo.: Rocky Mountain Petroleum Economics Institute, University of Wyoming, 1970).

Pushkarev, B. "Energy in the New York Region," pp. 13-23, in R.H. Connery and R.S. Gilmour (eds.), *The National Energy Problem* (Lexington, Mass.: D.C. Heath, Lexington Books, 1974).

Ramsey, F.P. "A Mathematical Theory of Saving," *Economic Journal*, 38 (1928), pp. 543-59.

Regional Plan Association, *Transportation and Economic Opportunity: A Report to the Transportation Administration of the City of New York* (New York, 1973).

Roberts, M.J. "Is There an Energy Crisis?," *Public Interest*, #31 (1973), pp. 17-37.

Rosenberg, N. "Innovative Responses to Materials Shortages," *American Economic Review*, Papers 63 (1973), pp. 111-18.

Ross, P.N. *Development of the Nuclear Energy Economy* (Pittsburgh: Westinghouse Electric Corporation, 1973).

Sagan, L.A. "Human Costs of Nuclear Power," *Science*, 177 (1972), pp. 487-93.

Schulze, W.D. "The Optimal Use of Non-Renewable Resources: The Theory of Extraction," *Journal of Environmental Economics and Management*, 1 (1974), pp. 53-73.

Schurr, S.H. (ed.). *Energy, Economic Growth and the Environment* (Baltimore: Johns Hopkins Press for Resources for the Future, 1973).

_____ , and P.T. Homan. *Middle Eastern Oil and the Western World: Prospects and Problems* (New York: American Elsevier, 1971).

Scott, A.D. *Natural Resources: The Economics of Conservation* (Toronto: University of Toronto Press, 1955).

_____ . "The Theory of the Mine under Conditions of Certainty," pp. 25-62, in M. Gaffney (ed.), *Extractive Resources and Taxation* (Madison: University of Wisconsin Press, 1967).

Scott, D.L. *Pollution in the Electric Power Industry: Its Control and Costs* (Lexington, Mass.: D.C. Heath, Lexington Books, 1973).

Smith, B.W. "Analysis of the Location of Coal-Fired Power Plants in the Eastern United States," *Economic Geography*, 49 (1973), pp. 243-50.

Smith, V.K. *Technical Change, Relative Prices and Environmental Resource Evaluation* (Baltimore: Johns Hopkins Press, 1974).

Smith, V.L. "Economics of Production from National Resources," *American Economic Review*, 58 (1968), pp. 409-31.

_____ . "An Optimistic Theory of Exhaustible Resources," *Journal of Economic Theory*, 9 (1974), pp. 384-96.

Solow, R.M. "The Economics of Resources and the Resources of Economics," *American Economic Review*, Papers 64 (1974), pp. 1-14.

Spivey, W.A., and W.A. Wecker. "Regional Economic Forecasting: Concepts and Methodology," *Papers, Regional Science Association*, 28 (1972), pp. 257-76.

Tilton, J.E. *United States Energy R and D Policy* (Washington, D.C.: Resources for the Future, 1974).

Tybout, R.A., and G.O.G. Löf. "Solar House Heating," *Natural Resources Journal*, 10 (1970), pp. 268-326.

United Nations, World Energy Supplies, 1961-70 (New York: United Nations, 1973).

US Bureau of the Census. *Historical Statistics of the United States, Colonial Times to 1957* (Washington, D.C.: US Government Printing Office, 1960).

_____. *Statistical Abstract of the United States, 1974* (Washington, D.C.: US Government Printing Office, 1974).

US Bureau of Mines. *Minerals Yearbook* (Washington, D.C.: US Government Printing Office, annual).

US Cabinet Task Force on Oil Import Control. *The Oil Import Question* (Washington, D.C.: US Government Printing Office, 1970).

US Department of Housing and Urban Development. *Residential Energy Consumption. Single Family Housing Final Report* (Washington, D.C.: HUD Contract #H-1654, March 1973).

US Department of the Interior, *Final Environmental Statement for the Prototype Oil Shale Leasing Program*, Vol. II *Energy Alternatives* (Washington, D.C.: USDI, 1973).

US Environmental Protection Agency, Office of Emergency Preparedness. *The Potential for Energy Conservation* (Washington, D.C.: EPA, 1972).

US Federal Energy Administration. *Project Independence Blueprint* (Washington, D.C.: US Government Printing Office, 1974).

US Federal Power Commission, *National Power Survey, 1964* (Washington, D.C.: US Government Printing Office, 1965).

_____. *National Power Survey, 1970* (Washington, D.C.: US Government Printing Office, 1971).

_____. *The Potential for Conversion of Oil-Fired and Gas-Fired Electric Generating Units to Use of Coal* (Washington, D.C.: 1973).

_____, National Power Survey: Task Force Report, *Environmental Research* (Washington, D.C.: January 1974).

US Geological Survey. *Outer Continental Shelf Statistics* (Washington, D.C.: June 1973).

US Office of Science and Technology, *Patterns of Energy Consumption in the United States* (Washington, D.C.: US Government Printing Office, 1972).

US Sulfur Oxide Control Technology Assessment Panel, *Final Report on Projected Utilization of Stack Gas Cleaning Systems by Steam Electric Plants* (Washington, D.C.: 1973).

Utton, A.E. (ed.). *National Petroleum Policy: A Critical Review* (Albuquerque: New Mexico University Press, 1970).

Vousden, N. "Basic Theoretical Issues of Resource Depletion," *Journal of Economic Theory*, 5 (1973), pp. 126-43.

Waverman, L. *Natural Gas and National Policy: A Linear Programming Model of North American Gas Flows* (Toronto: University of Toronto Press, 1973).

Wen, C.F., and R.T. Newcomb, "The Potential of Coal-Based Fuel and Energy Complexes: A Regional Appraisal" (National Science Foundation—RANN Report, January 1974).

White, D.E. "Geothermal Resources," in P. Kruger and C. Otte (eds.), *Geothermal Energy: Resources, Production, Stimulation* (Stanford: Stanford University Press, 1973).

Wildhorn, S., B.K. Burright, J. Enns, and T.F. Kirkwood. *How to Save Gasoline: Public Policy Alternatives to the Automobile* (Santa Monica: RAND Corporation, 1974).

Willrich, M., and T.B. Taylor, *Nuclear Theft: Risks and Safeguards* (Cambridge, Mass.: Ballinger, 1974).

Wilson, C.L. "A Plan for Energy Independence," *Foreign Affairs*, 51 (1973), pp. 657-75.

Wilson, C.L., and W.H. Matthews (eds.). *Man's Impact on the Global Environment* (Cambridge, Mass.: Study of Critical Environmental Problems, MIT, 1970).

Zupan, J.M. *The Distribution of Air Quality in the New York Region* (Baltimore: Johns Hopkins Press, 1973).

Index

Index

Accelerated supply, 7, 18, 19, 20; of coal, 73
Adelman, M.A., 66, 129, 130
Air-conditioning, 20, 139, 140
Air pollution, 152-54, 162
Alaska, 80, 116, 130, 162, 178, 193; gulf of, 162-63; and pipeline debate, 163-65
Algeria, 116
American Gas Association, 169
Appalachia, 73, 75, 76
Appliances, consumption of, 139-40
Arab-Israel War: of 1967, 65-118; of 1973, 120, 184
Arabs, 49, 120, 125, 129, 131, 137, 189; embargo, 121-23
ARCO, 103
Arizona, 106
Arrow, K.J., 49
Asia, 12
Atomic Energy Commission (AEC), 91-92, 94, 96, 97, 98, 99, 100, 101, 102, 157, 168, 169; forecasts, 155; record, 155
Automobiles: changes in stock, 137; compared to other modes, 141; controls on use of, 145, 146; embargo effects on, 121, 122; emissions, 162; fuel and, 141, 142, 149; and gasoline tax, 144-45; and horsepower tax, 143, 146-47; number of, 141; pollution, 162; and public transit, 142, 145; redesign of, 143, 168, 193

Backstop technology, 47-48; nuclear power as, 93-94
Balance of payments, 126, 183, 188
Banks, F.E., 39
Barnett, H.G., 34, 35, 36, 37
BART, 145
Baumol, W.J., 50
Bethe, H.A., 92
"Black lung," 76
Boulding, K.E., 147
Bradley, P.G., 39
Brannon, G.M., 172, 173
Breeder reactors, 24, 25, 32, 89, 90, 91, 92, 93, 94, 170, 196

Breyer, S.G., 174
British Petroleum, 119, 165
Brookings Institution, 134
Btu: conversion rates, 8n; standard of value, 179-81, weaknesses of, 180-81; value of western coal, 76
Bureau of Land Management, 73
Bureau of Mines, 75
Burnham, D.C., 6

Cabinet Task Force on Oil Imports Control (CTFOIC), 64, 65-66, 67
California, 76, 154, 157
Canada, 13, 65-66, 68n, 80, 81; and TAP, 163, 164-65
Cancer, 99
Carbon: dioxide, 32, 38, 154; hydro-, 152, 153; monoxide, 152
Catalytic converter, 162, 197
Chicago, 161, 168
Cicchetti, C.J., 164, 165
Clean Air Act, 71, 72, 177, 197
Clean fuels deficit problem, 73
Coal, 8, 20, 71-79; and British economy, 27-30; competition with oil, 71, 78; and environment, 158-61, 194-95; and federal government, 72-73; gasification, 76, 79, 96, 104-05; health and safety in mining, 75-76; imports, 14; investment in, 72, 79; leasing, 72-73; obstacles to, 72, 159; output, 71, 71n, 74, 78-79; ownership of, 72-73; projections, 78-79; and railways, 77-78; and regions, 73, 74; reserves, 73, 74, 75, 159; sales trends in, 71; strip (surface) mining, 71, 77, 79, 160, 161; and reclamation, 75; sulfur in, 73, 76, 154-55, 160, 161, 197, and transportation, 77-78; uncertainty of demand for, 78; and water, 77; western, 20, 73, 74, 76-77, 151, 161, 178, 194, 197
Coal Question, The, 27-30
Coal Mine Health and Safety Act, 75-76
Colorado, 103
Cole, H.S.D., 33

227

Concentration in oil industry, 62-63
Congress, 63, 66, 185
Conservation, 7-8, 20, 36, 50, 132; domestic, 138-40; in embargo, 122; in industry, 142-43; and life styles, 140-41; policies for, 143-49; savings from, 143, 148, 149, 180; in transportation, 141-42; and waste, 146-47. *See also* Demand restraint
Consumption: of electricity, 85, 139, 140; and embargo, 122; of energy, 8-11, 12, 15, 20, 146; and income distribution, 138n, 172; future, 53; of natural gas, 80; of oil, 115, 116, 130
Costs: coal, 71, 75, 79; in UK, 28; environmental, 147-48, 163; of extractive output, 35, 36, 42, 46; nuclear, 96, 97; oil, 58, 59, 64, 68, 120; shale, 103; solar power, 106-07; stockpile, 133-36; synthetics, 133-36
Crisis: impact of, 121-23, 199-200; reality of, xi, 184-85, 191
Cryogenic cables, 25, 157

Darmstadter, J., 4, 6
Dasgupta, P., 51
Deaths, 75-76, 99, 153, 156; of birds and fish, 162
Demand restraint, 7, 136, 137-49, 192-93; benefits of, 137-38; in the home, 138-40, 149; in industry, 142-43; policies for, 143-49; savings from, 143, 148, 149; in transportation, 141-42; and urban life styles, 140-41, 149; and waste, 146-47
Desulfurization, 73, 78, 160
Diesel engines, 151
Discount rate, 164; social and private, compared, 48-53; zero, 51, 52
Discoveries, 79-80, 81, 83
Dollar and oil prices, 118-119
Donora, 153
Drilling: geothermal, 108-09; of natural gas, 81, 83

Economic growth and energy, 3-8, 16
Elasticity: of demand, 6, 17, 29, 31, 137, 138; for electricity, 147; for gasoline, 144; of supply, of coal, 73, 160; of gas, 80, 83, 83n; of oil, 62
Electric Power Research Institute (EPRI), 170, 195
Electricity: and backwash technology, 47-48; bills, 85, 120; consumption in the home, 139, 140; and FPC, 87; generating capacity in, 85; nuclear, 94, 95; growth of, 85, 86n, 147, 178; in UK, 28; and National Power Surveys, 87; ownership of, 85, 87, 198; power plants, 85, 86, 143, 152-58; cutbacks in, 88; prices, 120, 138; reform of, 147-48, 190, 193; problems of, 195; projections, 15; rationalization of, 87; R and D, 149, 158; technical change and, 8, 87, 88, 143, 158; transmission, 87, 88, 157
Embargo, 65, 132, 134, 188; effects of, 121-23; and industry, 121-22
Emergency Petroleum Allocation Act, 122, 198
Enrichment plants, 97-98
Energy independence: costs of, 17, 186-89; psychology of, 189
Energy Research and Development Administration, 91n, 102, 169, 171, 196, 198
Energy stamps, 178
Environment, 151-65; and Alaskan oil, 163-65; and automobiles, 162; and coal, 73, 76-77, 158-61; costs, 147-48, 163, 165; and pricing, 147-48; damages tax, 44; and energy policy, 190-91; global, 34, 38, 154; in *Limits* model, 32; and oil spills, 162-63; and power plants, 152-58; and shale, 103-04; and synthetics, 104-05. *See also* Pollution
Environmental Protection Agency, 143, 155, 197, 198
Equilibrium, supply-demand, 16, 17
Europe, Western, 12, 13, 131, 132, 135
Exhaustible resources, theory of, 39-53, 126; and backstop technology, 47-48; decline in quality, 45-46; discount rates in, 48-53; profit-growth rule, 39-43; parameter changes in, 42; recycling in, 44-45; taxes in, 43-44, 51-52; environmental damages, 44; technological change in, 46-47
Expensing, 61
Exploration: in natural gas, 83; in oil, 62

Exponential growth, 28, 31, 32, 33
Exports: capacity of OPEC, 116; of fuels, 13, 14; Middle East, 130; world, 120
Extractive costs, 35, 36, 42; and technical change, 46

Federal Energy Administration, 22, 88, 135, 171, 198. *See also* Project Independence
Federal Power Act, 174
Federal Power Commission, 81, 185, 194, 195; and electricity, 87, 174, 175; pricing principles, 84; regulation of natural gas, 83-85, 174-75
"Floor-price" plan, 188
Ford Foundation Study, 5, 7; and coal, 79; policy proposals, 175, 177-79; projections, 6, 22-23; on solar power, 108
Foreign tax credits, 62
France, 23, 24, 91, 131, 170
Freeman, S.D., 159
Futures markets, 42, 47, 136
Futurology, 30

Gasoline: elasticity of demand for, 144; tax, 144-45, 186, 187
Geneva Agreements, 118-19
Geothermal energy, 108-09; dry-steam, 108; hot-water, 109; methods, 109; reserves, 109
Gordon, R.L., 39, 52
Government: involvement, 198-99; priorities, 199; and R and D, 168; reluctance to intervene, 199. *See also* Policy
Gray, L.C., 39
Green River Basin, 103
Gulf: of Alaska, 162-63; coast, 133

Half-life, 100n
Halvorsen, R., 147
Heal, G.M., 51
Health: and air pollution, 153, 155; in coalmining, 75-76; and nuclear power, 98-100
Heat rate, 15, 15n
Heller, W., 148
Herfindahl, O.C., 51
Hirshleifer, J.H., 49
Historical growth, 22-23, 25, 179

Horsepower statistics, 10-11
Hotelling, H., 39
Housing: change in stock, 137; and conservation, 138-39, 149
Houthakker, H.S., 144
HUD study, 138-39
Hudson-Jorgenson model, 6, 16, 123
Hudson River, 99, 156
Hydrocarbons, 152, 153
Hydroelectric power, 8, 9, 85, 154n
Hydrogen, 110-11; safety, 111; transmission, 111

IMF, 123
Imperial Valley, 108-109
Imports: from Canada, 68n; coal, 14; of fuel, 13, 14; natural gas, 14, 81; oil, 14, 18, 19, 57, 58, 134, 135, 176, 187; history of, 64-66; security and, 18n, 67-69; tariffs vs. quotas for, 66-67; uranium, 97, 102
Income distribution: and energy consumption, 138n, 177, 178-79; and energy prices, 138, 172
Indian Point, 99, 155, 156
Industry: conservation in, 142-43; oil, 62-63, 117-19, 125, 194, 198. *See also* Coal; Electricity; Gasoline; Nuclear Power
Inflation, and oil, 123, 125-26
Institute of Gas Technology, 169
Intergenerational equity, 51-52
International comparisons, 5-6
International Energy Program (IEP), 135, 176
Interstate Commerce Commission (ICC), 177
Inventories, 121, 133
Investment: in coal, 72, 79; in enrichment plants, 98; in nuclear power, 94, 98; requirements, 22
Iran, 49, 116, 117, 120, 124, 127, 128

Japan, 23, 116, 124, 131, 132, 135, 165, 170
Jevons, W.S., 27, 28, 29, 30, 34
Jones Act, 165

Khazzoom, J.D., 83
Kissinger, H.A., 67
Krutilla, J.V., 154n
Kuwait, 116, 118

Latin America, 12, 13
Lead times, 22, 72, 83, 102
Leasing: of coal, 72-73, 179; of natural gas, 83
Levy, W., 130
Libya, 49, 116, 117, 118
LMFBR. *See* Breeder reactors
Limits to Growth, The, 27, 31-34
Lind, R.C., 49
Location of power plants, 157-58, 198
Lovins, A.S., 97
Lurgi process, 104
Lyons, Kansas, 157

MacAvoy, P.W., 174
Magnetohydrodynamics (MHD), 158
Malthus, T.R., 28, 35, 36
Mandatory Oil Import Program, 66
Manhattan, 99, 141
Maximum efficient rate (MER), 59
McDonald, S.L., 60
Meadows, D.H., 32n
Meuse Valley, 153
Middle East, 68n, 130, 131, 132; dominance in oil, 115-16
Minerals Leasing Act, 72
MIT Study: and coal, 78; projections, 16-18; proposals, 175-77; stockpile estimates, 134, 136; and synthetics, 105
Montana-Wyoming Aqueduct Study, 77
Morse, C.W., 34, 35, 36, 37

National Coal Association, 161
National debt, 29
National Electric Reliability Council (NERC), 87
National Environmental Policy Act, 163, 197
National Petroleum Council (NPC), 22, 96, 108
Natural gas, 79-85, 194; advantages of, 79; bonus payments system, 83; consumption of, 80; drilling, 81, 83; elasticity of supply, 80, 83, 83n; and FPC, 83-85, 174-75; imports, 14, 81; interdependence with oil, 69, 84, 84-85n; prices, 80, 81, 82, 84; contract, 82, 84; regulated 83-85; rolled-in, 82, 83; well-head, 82; and regulation, 83-85, 174-75; reserves, 11, 79-80, 81; share of, 79; shortage, 80, 81-82; trade routes for, 80n; transport of, 80
New Mexico, 76
New York, 140, 142, 146, 153, 155, 156, 158; World Trade Center, 148
Nigeria, 116, 119
Nitrogen oxides, 152, 153, 154
Nordhaus, W.D., 36, 38, 39, 47, 96
North Sea, 130
North Slope, 163, 164
Nuclear power: accident risks, 155, 156; and AEC, 91-92, 94, 96, 97, 98, 99, 100, 101, 102; as backstop technology, 47, 93-94; breeder, 24, 25, 32, 89, 90, 91, 92, 93, 94, 170, 196; capacity, 23-24, 94, 95, 96; vs. coal, 99-100, 196; competitiveness of, 96; conclusions on, 101-02; costs, 96, 97; and deuterium, 92-93; and electricity, 85; enrichment plants, 97-98; fusion, 89, 89n, 92-93; health hazards of, 98-100; investment in, 94, 98; issues of, 89-90, 195-96; and plutonium, 90, 100n; pollution, 99-100; production, 94, 95, 96; projections, 23-24, 96, 155; R and D, 94, 96, 170; radiation, 99-100; radioactive waste disposal, 100-01, 157; reactors, 89, 90-92, 155; BWR, 90, 91, 99; PWR, 90, 91, 99; resources, 37; safety, 95, 99, 102, 156; theft risks, 101; and thorium, 90, 91; and tritium, 92; and uncertainty, 102; and uranium, 89, 90, 91, 95, 96-98; imports, 97, 102; milling, 94; prices, 97; reserves, 96-97
Nuclear Regulatory Commission, 91n, 102, 196
Nuclear theft, 101

Oak Ridge, 91, 98, 99
OECD, 144
Office of Science and Technology, 20n
Offshore production, 57, 58, 80, 162
Oil: and coal, 71; consumption, 115, 116, 130; costs, 58, 59, 64, 68, 120; embargo effects, 121-23; expensing, 61; export capacity, 116; foreign tax credits, 62; imports, 14, 19, 57, 58, 65, 121, 129, 134, 176, 187, 191-92; history of, 64-66; security

and, 67-69; tariffs and quotas on, 66-67, 68; industry and Congress, 63; and OPEC, 117-19, 125; and politics, 63; power of, 62-63, 194, 198; and MER, 59; and natural gas, 69, 84, 84-85n; offshore, 57; percentage depletion, 60-62, 63, 69; prices, 7, 68, 117, 118, 119-20, 136, 138, 161, 164, 186; production, 58, 115; prorationing, 59; regulation, 69; reserves, 11, 58, 115-16, 136; rule of capture, 58-59; spills, 151, 162-63; tariff, 65, 186; and quotas, 66-67, 68; storage option, 133; and taxes, 186, 187; taxes, 60-62, 186; and oil producers, 117, 118; trends in, 57, 58; unitization, 59-60; well-spacing, 58-59. *See also* Shale

Olson, C.E., 87

OPEC, 63, 64, 68n, 137, 138, 176, 179, 184, 187, 188, 192, 196; ability to absorb capital, 128; as a cartel, 124-25, 184; and consuming countries, 131-32; dangers for, 127-28; disturbances to monetary system, 124; domestic economies, 128; export capacity, 116, General Agreement on Participation, 119; heterogeneity of, 126-27; history of, 117-19; influence on inflation, 123, 125-26; justification for price increases, 125-26; and oil companies, 117-18, 125; strategies to deal with, 129-32; and theory of resource extraction, 126; and world recession, 124

Oppenheimer, B.I., 63

Optimists, 34-38

Particulates, 152-153
Passamaquoddy Bay, 110
Percentage depletion, 60-62, 63, 69
Permian Basin, 84
Pessimists, 27-34, 47
Petroleum, 57; inventories, 133; trade in, 14. *See also* Oil
Phillips Petroleum Co. v. Wisconsin, 84, 174
Photovoltaic conversion, 107
Plutonium, 90, 100n, 101, 195
Policy, 183-99; acceptable, 189-90; conservation and demand restraint, 143-49. 192-93; to deal with OPEC, 129-32; diagnosis, 183-86; and energy independence, 186-89; and energy supply, 193-97; and environment, 190-91, 197-98; and FPC, 174-75; incremental, 190; instruments, 143-49, 193, 199; issues, 167-81, 185; no action, 189; objectives, 183-84; and oil imports, 191-92; proposals, 189-99; Ford Foundation, 175, 177-79; MIT, 175, 176-77; Project Independence, 175-76; R and D, 167-72; stockpiles, 132-36, 176, 188, 192; and taxes, 172-75

Politics and oil industry, 63

Pollution: air, 152-54, 162; and Alaskan oil, 163-65; and animal life, 162, 164; automobile, 162; global, 34, 38, 154; impacts, 151-52; in *Limits* model, 32, 34; in Nordhaus model, 37-38; ocean, 151, 162-63; PIB estimates of, 151-52; power plant, 152-58; standards, 154-55; taxes, 173; thermal, 32, 151, 156-57; water, 75, 77, 151, 152, 156. *See also* Environment

Power plants, 78, 85, 86, 151, 159; and environment, 152-58; land use demands of, 157; location of, 157-58, 198; nuclear, 23-24, 96, 99; R and D in, 158; and technical progress, 87, 88, 143, 158

Price-Anderson Act, 155

Prices: coal, 71; electricity, 120, 138, 147-48, 190, 193; energy, 16, 63, 138, 172; gasoline, 120; natural gas, 80, 81, 82; regulated, 83-85, 174-75; oil, 7, 68, 117, 118, 119-20, 136, 138, 161, 164, 186; recourse, 36, 46; shadow, 37; uranium, 97, 102

Prince William Sound, 164

Production potential, 20, 22

Profit-growth rule, 39-43; parameter changes in, 42

Project Independence: and coal, 78; and energy consumption, 149; policy proposals, 6-7, 175-76; and pollution impacts, 151-52; projections, 6, 18-22; and solar power, 107, 108; stockpile estimates, 135-36; and synthetics, 105

Projections, 15-26; coal, 78-79; of

Projections (cont.))
  Jevons, 27-30; compared, 16-23; and embargo, 121, 122-23; Ford Foundation, 22-23; long-term, 25-26; MIT study, 16-18; of Nordhaus, 37; nuclear power, 23-24, 96, 155; pitfalls, 15-16; Project Independence, 18-22; technological, 25
Prorationing, 59
Prudhoe, 163, 164

Quotas, 66, 133; v. tariffs, 66-67, 68

Radiation: exposure, 99; risks, 155, 156
Radioactive wastes, disposal of, 100-01, 157, 195-96
Railways: and coal, 77-78; efficiency of, 141, 142
Ramsey, F.P., 51, 52
Rance Estuary, 110
RAND, 144
Rankine Cycle, 158
Ras Tanura, 120
Rate: of extraction, 39, 42-43, 44; of interest, 40, 42, 43, 46, 48, 134; zero, 52. *See also* Discount rate
Reclamation costs, 75, 161
Recycling, 44-45
Regions and coal, 73, 75, 76-77
Regulation: of electricity, 87; of gas, 83-85, 174-75; of oil industry, 69
Research and development: in electric utilities, 149, 158; estimates, 169, 170, 171; evaluation of, 167-72, 190, 196-97, 199; financing of, 171; government and, 168; nuclear power, 170, 190; optimal level of, 167-68, 169, 171; organization of, 171-72; and policy goals, 168-69; subsidies, 173
Reserves, 11-12, 48; of coal, 73, 74, 75, 159; -consumption ratios, 31, 37; geothermal, 109; natural gas, 79, 80, 81; -production ratios, 80-81; oil, 11, 58, 115-16, 136; shale, 103; uranium, 96-97
Residential and commercial energy savings, 138-41
Resource: decline in quality, 45-46; scarcity, in *Limits* model, 31-34, Malthusian and Ricardian, 35, 36

Ricardo, D., 35, 36
Risks, in theory of resource extraction, 49
Roberts, M.J., 7
Royalties, 41, 42-43, 45-46, 48, 62, 117
Rule of capture, 58-59
Rural Electrification Administration (REA), 85, 87

Safety: of hydrogen, 111; nuclear, 95, 98-102, 156
Santa Barbara, 162
Saudi Arabia, 13, 49, 116, 117, 120, 124, 130
Schulze, W.D., 39
Scott, A.D., 39, 49
Security, 67-69
Shale oil, 103-04; costs, 103; impact on environment, 103-04; reserves, 11, 103; water requirements, 103
Shapiro, D.L., 49
Shell, 119
Social saving program, 53
Solar power, 106-08; advantages and disadvantages, 106; costs, 106-07, 208; forecasts of, 108; methods, 107; scope for, 107-08
Solow, R.M., 39, 51, 52
State Department, 188
Stationary state, 30, 52
Stock of resources, 41, 42
Stockpiles, 18, 132-36, 176, 188, 192; costs, 133-36; disadvantages of, 176; estimates of, Brookings Institution, 134; MIT Study, 134, 136; Project Independence, 135-36; and GNP tradeoffs, 134, 135; methods of, 133
Strip (surface) mining, 71, 77, 79, 160, 161; and environment, 73, 75, 76-77, 159, 160, 161; reclamation costs, 75, 161
Subsidies for R and D, 173
Suez Canal, 65, 118
Sulfur: in coal, 73, 76, 154-55, 161, 197; oxides, 152, 153, 154; standards, 154-55, 160
Supply: elasticity of, for coal, 73, 160; for gas, 80, 83, 83n; for oil, 62; interruptions, 65, 68
Supreme Court, 57, 84, 174, 193
Surface Mines Association, 75

Sweden, 23, 24, 97
Synthetic fuels, 22, 77, 78, 104-05; costs of, 105; prospects for, 105
System dynamics, 31-34

Take-off dates, 25
Tankers, 162
Tar sands, 11
Tariffs, 65, 66-67, 68, 133; or taxes, 186, 187
Taxes: annuity, 43; corporate, 50; distortions of, 50, 173; and energy policy, 172-74; environmental damages, 44; and oil producers, 117, 118; of petroleum, 60-62, and neutrality, 61-62; or tariffs, 186, 187; uniform, 43
Technical Fix, 22-23, 25, 177
Technical progress: in electricity, 8, 87, 88, 143, 158; and natural resources, 34, 35-36; in theory, 46-47, 52
Tehran Agreement, 118, 119, 184
Tennessee Valley Authority (TVA), 85, 87
Thermal efficiency, 8, 15, 16, 91, 143n, 149
Tidal power, 110
Time preference, 51
Torrey Canyon, 162
Trade: in fuels, 13, 14; in natural gas, 80n, 81. *See also* Exports; Imports
Trans-Alaska Pipeline, 163-65
Trans-Arabian Pipeline, 65, 118
Trans-Canadian Pipeline, 164, 165
Transmission: of electricity, 25, 87, 88, 157; of hydrogen, 111
Transportation, 4; of coal, 77-78; energy conservation in, 141-42; of natural gas, 80; and pollution, 151; public, 142, 145; waste in, 146. *See also* Automobiles; Railways
Trends, 8-14; consumption, 8-11, 12; in oil, 57, 58; in trade, 13, 14

Tripoli Agreements, 118, 119
Tsunamis, 162

Uncertainty: of demand for coal, 78; and nuclear power, 102; in theory, 49-50
Unemployment, 122, 144
United Kingdom, 23, 24, 76, 91, 94, 170
Unitization, 59-60
Uranium, 89, 90, 91, 95, 99, 102; enrichment plants, 97-98; imports, 97, 102; milling, 94; prices, 97, 102; reserves, 96-97
Urban lifestyles and energy, 140-41, 149
United States Department of Interior, 25, 164
US Geological Survey, 109, 161
USSR, 12, 23, 24, 91, 115, 116

Valdez, 163
Venezuela, 13, 68n, 81, 115, 116, 117, 118, 120
Verleger, P., 144n

Water: and coal, 77; pollution, 75, 77, 151, 152, 156; and shale, 103
Water Pollution Control Act, 151
Waverman, L., 80n
Ways and Means Committee, 187
Well-spacing, 58-59
West Germany, 23, 24, 170
West Valley, 156
Windmills, 107
World: exports, 120; nuclear projections, 23-24; trade, 12-13; and US compared, 12, 137
Wright-Pastore scheme, 187

Zero energy growth, 22-23, 177
Zero rate of interest and extraction, 52

About the Author

Harry W. Richardson is a professor of economics at the University of Southern California. He received his undergraduate and graduate education at the University of Manchester, England, and held appointments in several British universities and at the University of Pittsburgh. His earlier books include *Economic Recovery in Britain, 1932-9* (1967), *Building in the British Economy between the Wars* (with D.H. Aldcroft, 1968), *Regional Economics: Location Theory, Urban Structure and Regional Change* (1969), *The British Economy, 1870-1939* (with D.H. Aldcroft, 1969), *Elements of Regional Economics* (1969), *Regional Economics: A Reader* (editor, 1970), *Urban Economics* (1971), *Input-Output and Regional Economics* (1972), *Regional Growth Theory* (1973), *The Economics of Urban Size* (1973), *Regional Policy and Planning in Spain* (1975), and (with others) *Housing and Urban Spatial Structure: Twentieth Century Edinburgh* (1975).